TSUNAMI AND SEICHE HAZARDS IN SITE EVALUATION FOR NUCLEAR INSTALLATIONS

The following States are Members of the International Atomic Energy Agency:

AFGHANISTAN
ALBANIA
ALGERIA
ANGOLA
ANTIGUA AND BARBUDA
ARGENTINA
ARMENIA
AUSTRALIA
AUSTRIA
AZERBAIJAN
BAHAMAS
BAHRAIN
BANGLADESH
BARBADOS
BELARUS
BELGIUM
BELIZE
BENIN
BOLIVIA, PLURINATIONAL STATE OF
BOSNIA AND HERZEGOVINA
BOTSWANA
BRAZIL
BRUNEI DARUSSALAM
BULGARIA
BURKINA FASO
BURUNDI
CABO VERDE
CAMBODIA
CAMEROON
CANADA
CENTRAL AFRICAN REPUBLIC
CHAD
CHILE
CHINA
COLOMBIA
COMOROS
CONGO
COOK ISLANDS
COSTA RICA
CÔTE D'IVOIRE
CROATIA
CUBA
CYPRUS
CZECH REPUBLIC
DEMOCRATIC REPUBLIC OF THE CONGO
DENMARK
DJIBOUTI
DOMINICA
DOMINICAN REPUBLIC
ECUADOR
EGYPT
EL SALVADOR
ERITREA
ESTONIA
ESWATINI
ETHIOPIA
FIJI
FINLAND
FRANCE
GABON
GAMBIA, THE
GEORGIA
GERMANY
GHANA
GREECE
GRENADA
GUATEMALA
GUINEA
GUYANA
HAITI
HOLY SEE
HONDURAS
HUNGARY
ICELAND
INDIA
INDONESIA
IRAN, ISLAMIC REPUBLIC OF
IRAQ
IRELAND
ISRAEL
ITALY
JAMAICA
JAPAN
JORDAN
KAZAKHSTAN
KENYA
KOREA, REPUBLIC OF
KUWAIT
KYRGYZSTAN
LAO PEOPLE'S DEMOCRATIC REPUBLIC
LATVIA
LEBANON
LESOTHO
LIBERIA
LIBYA
LIECHTENSTEIN
LITHUANIA
LUXEMBOURG
MADAGASCAR
MALAWI
MALAYSIA
MALI
MALTA
MARSHALL ISLANDS
MAURITANIA
MAURITIUS
MEXICO
MONACO
MONGOLIA
MONTENEGRO
MOROCCO
MOZAMBIQUE
MYANMAR
NAMIBIA
NEPAL
NETHERLANDS, KINGDOM OF THE
NEW ZEALAND
NICARAGUA
NIGER
NIGERIA
NORTH MACEDONIA
NORWAY
OMAN
PAKISTAN
PALAU
PANAMA
PAPUA NEW GUINEA
PARAGUAY
PERU
PHILIPPINES
POLAND
PORTUGAL
QATAR
REPUBLIC OF MOLDOVA
ROMANIA
RUSSIAN FEDERATION
RWANDA
SAINT KITTS AND NEVIS
SAINT LUCIA
SAINT VINCENT AND THE GRENADINES
SAMOA
SAN MARINO
SAUDI ARABIA
SENEGAL
SERBIA
SEYCHELLES
SIERRA LEONE
SINGAPORE
SLOVAKIA
SLOVENIA
SOMALIA
SOUTH AFRICA
SPAIN
SRI LANKA
SUDAN
SWEDEN
SWITZERLAND
SYRIAN ARAB REPUBLIC
TAJIKISTAN
THAILAND
TOGO
TONGA
TRINIDAD AND TOBAGO
TUNISIA
TÜRKİYE
TURKMENISTAN
UGANDA
UKRAINE
UNITED ARAB EMIRATES
UNITED KINGDOM OF GREAT BRITAIN AND NORTHERN IRELAND
UNITED REPUBLIC OF TANZANIA
UNITED STATES OF AMERICA
URUGUAY
UZBEKISTAN
VANUATU
VENEZUELA, BOLIVARIAN REPUBLIC OF
VIET NAM
YEMEN
ZAMBIA
ZIMBABWE

The Agency's Statute was approved on 23 October 1956 by the Conference on the Statute of the IAEA held at United Nations Headquarters, New York; it entered into force on 29 July 1957. The Headquarters of the Agency are situated in Vienna. Its principal objective is "to accelerate and enlarge the contribution of atomic energy to peace, health and prosperity throughout the world".

SAFETY REPORTS SERIES No. 116

TSUNAMI AND SEICHE HAZARDS IN SITE EVALUATION FOR NUCLEAR INSTALLATIONS

INTERNATIONAL ATOMIC ENERGY AGENCY
VIENNA, 2025

Publishing Section
International Atomic Energy Agency
Vienna International Centre
PO Box 100
1400 Vienna, Austria
tel.: +43 1 2600 22529 or 22530
email: sales.publications@iaea.org
www.iaea.org/publications

Printed by the IAEA in Austria
March 2025
STI/PUB/1988
https://doi.org/10.61092/iaea.dkpf-b3tq

IAEA Library Cataloguing in Publication Data

Names: International Atomic Energy Agency.
Title: Tsunami and seiche hazards in site evaluation for nuclear installations / International Atomic Energy Agency.
Description: Vienna : International Atomic Energy Agency, 2025. | Series: IAEA safety reports series, ISSN 1020–6450 ; no. 116 | Includes bibliographical references.
Identifiers: IAEAL 24-01718 | ISBN 978-92-0-133424-4 (paperback : alk. paper) | ISBN 978-92-0-133624-8 (pdf) | ISBN 978-92-0-133524-1 (epub)
Subjects: LCSH: Nuclear facilities — Safety measures. | Nuclear power plants — Safety measures. | Natural disasters. | Tsunamis.
Classification: UDC 621.039.58 | STI/PUB/1988

FOREWORD

IAEA Safety Standards Series No. SSG-18, Meteorological and Hydrological Hazards in Site Evaluation for Nuclear Installations, was published in 2011 and provides evaluation methodologies for meteorological and hydrological hazards with guidance for their application.

SSG-18 describes all flooding hazards for nuclear installations. Since tsunamis and seiches can be major flooding hazards for nuclear sites located on the coast, this Safety Report supports the specific areas of SSG-18 relating to these and provides detailed methodologies and case studies that can be used by Member States for the evaluation of tsunami and seiche hazards.

Since the Great East Japan earthquake and tsunami of 11 March 2011, the importance of protecting nuclear facilities against coastal flooding and, in particular, tsunamis has received renewed attention. This event highlighted the need for deeper consideration of tsunami hazards in the design and safety assessment of nuclear installations, especially in terms of providing some defence in depth measures in such conditions.

In support of SSG-18, this Safety Report details state of the art methodologies and provides recent case studies that may be used for the evaluation of tsunami and seiche hazards.

The IAEA would like to thank all those who contributed to the drafting and review of this report. The IAEA officers responsible for this publication were Y. Iwabuchi and S. Nomura of the Division of Nuclear Installation Safety.

CONTENTS

1. INTRODUCTION

1.1. BACKGROUND

Some nuclear power plants in Kalpakkam, India, were affected by the Indian Ocean tsunami on 26 December 2004. As a result of this incident, the IAEA initiated the development of IAEA Specific Safety Guide No. SSG-18, Meteorological and Hydrological Hazards in Site Evaluation for Nuclear Installations [1]. SSG-18 was published in 2011 and provides evaluation methodologies for hydrological and meteorological hazards and guidance for their application. SSG-18 describes all flooding hazards, but since tsunamis and seiches are major flooding hazards for nuclear sites located on the coast, there was a need for a Safety Report that addressed those specific hazards. This publication therefore provides detailed methodologies and case studies that may be used by Member States for the evaluation of tsunami and seiche hazards.

In light of the Great East Japan Earthquake and Tsunami of 11 March 2011, the importance of protecting nuclear facilities against coastal flooding and, in particular, tsunamis has received renewed attention. The concurrent event of an earthquake followed by a tsunami caused the first severe accident at a nuclear power plant in the world involving multiple external hazards. As highlighted by the Fukushima Daiichi Accident — Report by the Director General [2], published by the IAEA in 2015, this event highlighted the need for deeper consideration of tsunami hazards in the design and safety assessment for nuclear installations, especially in view of providing some defence in depth measures in such conditions. Consequently, site evaluation and design methodologies are required to consider comprehensively the combined effects of earthquakes and tsunamis.

Records of extreme events, such as past large tsunamis, are either very sparsely documented or simply not available. There is a need to share the latest knowledge related to the evaluation of tsunami hazards and the effects of tsunamis for coastal nuclear sites and facilities from the experience gained from recent major tsunamis.

1.2. OBJECTIVE

The objective of this Safety Report is to detail state of the art methodologies and provide recent case studies that may be used for the evaluation of tsunami and seiche hazards as a supporting document to SSG-18.

Guidance and recommendations provided here in relation to identified good practices represent expert opinion but are not made on the basis of a consensus of all Member States.

1.3. SCOPE

The scope of this Safety Report covers the assessment of tsunami and seiche hazards during site evaluation and re-evaluation for a nuclear facility's design and safety assessment.

The aspects of the evaluation of tsunami and seiche hazards discussed are the following:

— Data collection;
— Specification and parameterization of sources;
— Propagation analysis;
— Hazard assessment of coastal impacts;
— Design and evaluation parameters;
— Management systems;
— Documentation.

1.4. STRUCTURE

Sections 2 and 3 present a general description of tsunamis and seiches, including the definition of terms, their generation mechanisms and the consideration of tsunami and seiche effects. Section 4 discusses the effects on nuclear sites and facilities of recent major tsunamis, such as the Indian Ocean tsunami in 2004 and the Great East Japan earthquake and tsunami in 2011. Section 5 presents an overview of the evaluation of tsunami and seiche hazards, as well as the uncertainties and concepts of deterministic and probabilistic approaches. Sections 6 to 8 provide detailed methodologies for the evaluation of tsunami and seiche hazards, such as data collection, the specification and parameterization of tsunami and seiche source models and propagation analyses. Section 9 provides a detailed assessment methodology for tsunami and seiche hazards and coastal impacts from the perspective of deterministic and probabilistic approaches on the basis of the detailed methodologies provided in Sections 6 to 8. Section 10 provides detailed methodologies for tsunami load effects to nuclear facilities on the basis of the evaluated tsunami and seiche hazards. The management aspects of these activities are discussed in Section 11, and Section 12 presents the recommended structure for documentation of the evaluation of tsunami and seiche hazards. The publication is completed by the Annex, which gives three examples of developed databases.

2. GENERAL DESCRIPTION OF TSUNAMIS

2.1. DEFINITION OF TERMS

2.1.1. Definition of a tsunami

A tsunami is a series of travelling waves of long wavelength (from a few kilometres to hundreds of kilometres) and period (from several minutes to tens of minutes and, exceptionally, hours), which are generated by deformation or disturbances of the seafloor (or, in generic terms, underwater floor). Earthquakes, volcanic phenomena, underwater and coastal landslides, rock falls or cliff failures can generate tsunamis. Large meteorites may also impact the ocean and generate a tsunami. Landslides can include submarine and subaerial slides and ice falls. Volcanic sources include the effects of pyroclastic flows, caldera collapses and submarine explosions. All oceanic regions and sea basins of the world, and even fjords or large lakes, can be affected by tsunamis.

2.1.2. Parameters for tsunami hazard assessment

The tsunami height is usually measured from sea level or datum, but the flow depth is measured from ground level. The horizontal distance of water penetration measured from the coast is called the 'inundation distance'. The height at maximum inundation, where the velocity and kinetic energy of a tsunami becomes zero, is called the 'runup' (see Fig. 1).

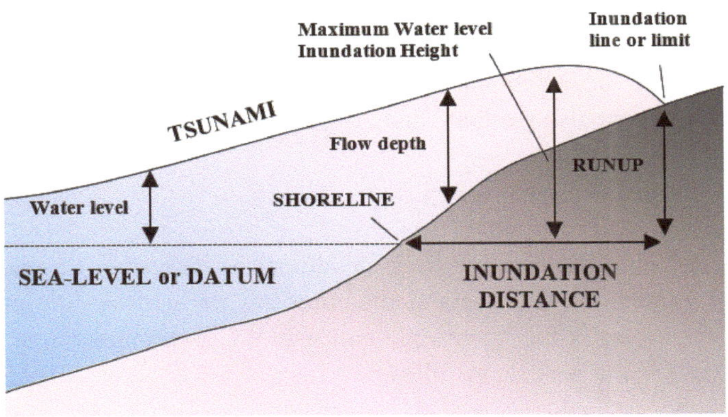

FIG. 1. Definitions of tsunami heights.

The results of hazard assessment for tsunami flooding should be the bounding values for the maximum water level at the shoreline, runup, flow depth, horizontal inundation distance, maximum water level at the plant site, minimum water level at the shoreline, current velocity, hydrodynamic force, duration of the drawdown below the intake and sediment transport.

2.1.3. Types of tsunami

2.1.3.1. Classification by distance to the source

SSG-18 [1] describes how tsunamis can be classified according to the distance across which they have an effect:

> "Tsunamis can also be classified as 'local' tsunamis or 'distant' tsunamis. A tsunami is called a local tsunami when it affects only the region near its source. Local tsunamis can be generated by earthquakes, volcanic activity and landslides. Earthquake induced local tsunamis represent the most frequent type of destructive tsunami. Less frequent but affecting wider regions are ocean wide or distant tsunamis that arrive at places remote from their source after travelling across the ocean or sea basins. Examples of destructive earthquake induced distant tsunamis include the 1960 Chilean tsunami, which affected many States around the Pacific Ocean, and the highly destructive 2004 Indian Ocean tsunami." (Ref. [1], para. 5.43).

The 2004 Indian Ocean tsunami affected many States around the Indian Ocean, including on the African coasts. The Great East Japan earthquake and tsunami of 2011 affected local nuclear power plant sites, including the Fukushima Daiichi nuclear power plant, but the tsunami also caused damage across the Pacific Ocean. Volcanic activities such as those associated with the 1883 Krakatau event also generate distant tsunamis.

2.1.3.2. Classification by tsunami size

Some tsunami scales have been developed to describe the magnitude or intensity of a tsunami. The tsunami magnitude is defined by Iida et al. [3] as $m = log_2 h$, where m is the magnitude and h is the maximum runup in metres. Hatori [4] revised the m scale as follows:

$$m = 2.7\log H + 2.7\log\Delta - 4.3 \quad (1)$$

where H is the runup in metres and Δ is the distance from source to coast in kilometres. Soloviev and Go [5] also defined the tsunami intensity (I) as

$$I = log_2(2^{1/2}h_0)$$

where h_0 is the average runup height of the wave. Figure 2 shows the distribution of historical tsunami events with a tsunami magnitude [3] estimated by the maximum water heights in National Centers for Environmental Information World Data Service for Geophysics (NCEI/WDS) Tsunami Event Database.

2.1.3.3. *Classification by coastal effects*

A tsunami can cause the inundation of land when its wavelength is so large that a huge mass of water follows behind the wavefront at high speed. The penetration of a tsunami can be classified into the following types based on the observation of the Great East Japan tsunami of 2011; these are shown in Fig. 3:

(a) Sea level change type;
(b) Overflow type;
(c) Bore type (breaking or broken wave);
(d) Runup type.

In cases where a tsunami has a long wavelength, the tsunami appears as a gradual sea level change, as shown in Fig. 3(a), which was seen at the Onagawa nuclear power plant in the Great East Japan earthquake and tsunami 2011.

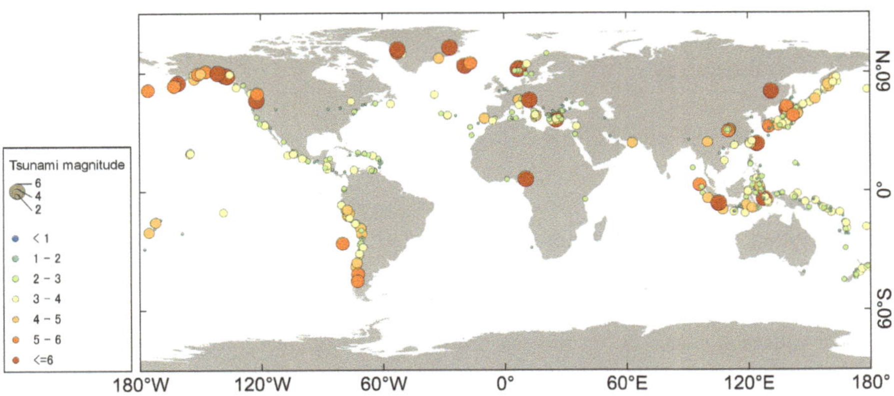

FIG. 2. Historical tsunami events from the National Centers for Environmental Information (NCEI)/World Data System (WDS) Tsunami Event Database (see Section A–1.1). Only those events with validity ≥ 0 are shown.

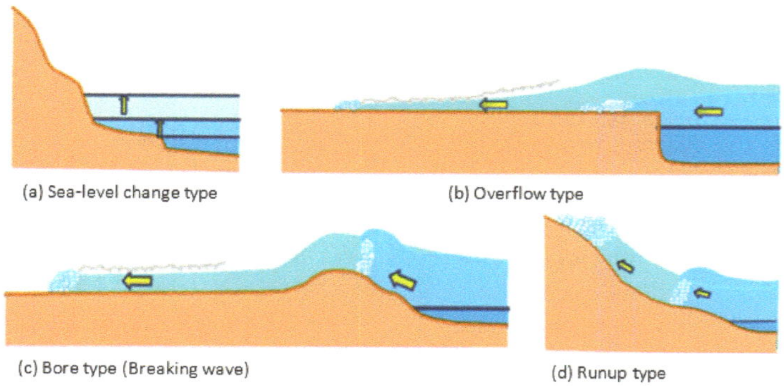

FIG. 3. Types of tsunami penetration in the 2011 Great East Japan earthquake and tsunami. Figure modified from the original. Reproduced from Ref. [6] with permission courtesy of Port and Airport Research Institute.

Figure 3(b) shows an overflow type, which is a tsunami that overtops the sea wall or sea bank with strong current velocity. The bore type, shown in Fig. 3(c), is a tsunami propagating with breaking at the front of the wave as a tsunami bore. When the slope of the coastal land area is steep, a tsunami breaking on the shore runs strongly up the slope, as shown in Fig. 3(d). These types of tsunami could cause strong currents and hydrodynamic force, which may exceed the hydrostatic pressure.

2.2. TSUNAMI GENERATION MECHANISMS

2.2.1. Earthquakes

SSG-18 [1] discusses how tsunamis can be generated by earthquakes:

"Earthquakes are the most frequent source of tsunamis. An earthquake induced tsunami is generated by a seafloor deformation associated with submarine and near-coast earthquakes with shallow depth (<50 km), large magnitude (M > 6.5) and dip–slip mechanism. Strike–slip fault motion produces a small vertical deformation of the seafloor, and consequently the induced tsunamis are usually of smaller height." (Ref. [1], para. 5.40).

2.2.2. Landslides

Tsunamis may also be triggered by landslides:

"Underwater and coastal (subaerial or subaerial–underwater) landslides, rock falls and cliff failures may also generate tsunamis, some of which are locally more disastrous than earthquake induced tsunamis. These landslides may or may not be triggered by an earthquake or by volcanic activity." (Ref. [1], para. 5.42).

2.2.3. Volcanic activities

Another significant cause of tsunamis is volcanic activity:

"Tsunamis may be generated by volcanic phenomena when voluminous (e.g. from 10^6 to greater than 10^9 m^3) landslides, pyroclastic flows or debris avalanches rapidly enter the sea or large lakes, or by the eruption of underwater volcanoes. Collapse of a volcano edifice triggered by a volcanic eruption or an earthquake may lead to large displacement of the slopes, which in turn can cause tsunamis in proximal bodies of water." (Ref. [1], para. 5.41).

2.2.4. Impact tsunamis

A gigantic meteorite impact occurred at the Yucatan Peninsula c. 65 Ma ago and resulted in the formation of the Chicxulub crater (see e.g. Ref. [7]). The meteorite impact may have caused massive tsunamis with heights of more than 200 m that reached the coastal area of North America [8]. For meteorite induced tsunamis, however, assessments conducted to date do not demonstrate that the frequency of occurrence exceeds the screening level usually adopted by Member States (see SSG-18 [1], footnote 32). Therefore, impact tsunamis will not be further considered in this document.

2.3. WAVE CHARACTERISTICS OF TSUNAMIS

2.3.1. Wave speed of tsunamis

The wave speed of a tsunami depends on the depth of the ocean. Since tsunami waves have wavelengths much longer than the depth of the ocean, they are classified as long waves with long periods. Figure 4 shows tsunami waves

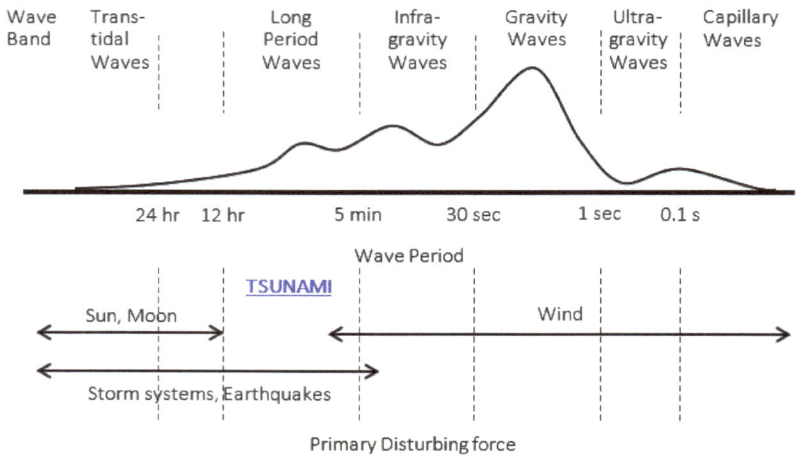

FIG. 4. Schematic representation of the energy contained in the waves of the ocean. Modified from http://co-ops.nos.noaa.gov/levelhow.html. Reproduced courtesy of National Oceanic and Atmospheric Administration.

compared with other ocean waves in the wave power spectrum. The speed of a tsunami is given by $\sqrt{gd}$, where g is gravitational acceleration and d is water depth. The wave speed of a tsunami travelling across the Pacific Ocean reaches the speed of a jet airplane. On reaching the coast, as described in SSG-18 [1], para. 5.38, "Owing to the fact that wave speed is reduced and wave length is shortened when the depth decreases, tsunami waves become steeper and increase in height on approaching shallow water." Thus, when the tsunami waves reach the coastal zone, they produce hazardous effects near and on the shoreline and can also produce hazardous effects far inland, depending on the coastal zone topography.

2.3.2. Wave transformation by coastal effects

> "In the coastal zone, local topography and bathymetry, such as a peninsula or submarine canyon, may cause an additional increase in wave heights. The wave heights could also be amplified by the presence of a bay, an estuary, a harbour or lagoon funnels as the tsunami moves inland." (Ref. [1], para. 5.38).

A recent study [9] identified tsunami-focusing phenomena due to an initial shape of the waveform, which might enhance runup substantially.

In a leading elevation wave, the wave crest arrives first. However, in the leading depression wave, the wave trough arrives first, which is characterized by a drawdown of water and retreat from the shore before a rise in the water level occurs. The characteristics of leading elevation or depression waves are related to fault displacement at the ocean floor [10, 11].

3. GENERAL DESCRIPTION OF SEICHES

3.1. CHARACTERISTICS OF SEICHES

Seiches are long period standing oscillations in an enclosed basin or a locally isolated part of a basin [12]. Abnormal oscillations of the water level occur with a period of approximately a few minutes to a few tens of minutes, depending on the forcing energy input to the basin and the shape and bathymetry of the basin. The amplitude of the forcing fluctuations may be anything from a few tens of centimetres up to around 2 m [13].

Harbour oscillations (coastal seiches) are a specific type of seiche motion that occurs in partially enclosed basins (bays, fjords, inlets and harbours) that are connected through one or more openings to the sea. Harbour oscillations differ from seiches in closed water bodies [12].

Seiche oscillations are also common in marginal seas (partly enclosed by land and not as deep as the open ocean) and may have an effect on small craft inside enclosed bays and harbours. Seiches can be dangerous for navigation and have an effect on port management, since they increase not only the effects of storm surges or tsunamis but also the influence of moderate storms inside harbours or bays. A recent study [14] reported that tidal, seiche and wind dynamics were observed in a lagoon in the Mediterranean Sea. According to the results of the obtained observations and simulations, the lagoon circulation was dominated by tidal, atmospheric and seiche forcing [14]. Hence, the monitoring of seiches in semi-enclosed basins, determining the general characteristics of seiche behaviour in the basin and assessing the spatial and temporal distribution of seiche amplitudes, for example, are important and worthy of further investigation.

Seiches are caused by short duration, rapid changes in atmospheric pressure with the passage of low or high pressure weather systems or by wave input to enclosed or semi-enclosed basins, such as from tsunamis. Seiches develop on great lakes, in other large, confined water bodies and semi-enclosed bays, as well as at continental shelves.

The height and duration of seiches are dependent on their forcing mechanisms, which are in general large scale spatial changes in the atmospheric pressure, asymmetric wind shear, etc. The agitation inside the basins depends on (i) the period of the waves forcing the agitation, (ii) the reflection and dissipation characteristics of the boundaries and (iii) the geometric properties of the basins. Waves (short period waves or tsunamis) continuously entering the basins, such as in the case of semi-enclosed bays, may in some cases cause abnormal water surface fluctuations and long period amplifications if their period coincides with one of the periods of free oscillations of the basin.

Extremely strong seiche oscillations are regularly observed in specific sea areas around the world. Certain harbours and ports are known to have frequent strong periodic horizontal water motions, such as those in Cape Town (South Africa), Los Angeles (USA), Dakar (Senegal), Toulon and Marseilles (France), Algiers (Algeria), Tuapse and Sochi (Russia), Batumi (Georgia) and Esperance (Australia). Seiche motions in these basins create unacceptable vessel movement, which can, in turn, lead to the breaking of mooring lines, fenders and piles and the onset of large amplitude ship oscillations and damage [12]. Nagasaki Bay, Japan, Longkou Harbour, China, and the Caribbean coast of Puerto Rico are other examples [15].

3.2. GENERATION MECHANISMS OF SEICHES

Seiches are generated by a wide variety of mechanisms, including tsunamis, seismic waves, jet-like currents and meteorological factors. Together with tsunamis and internal waves, the main sources of background long waves in the ocean are atmospheric processes. There are three major mechanisms that transfer the energy of atmospheric processes into long waves in the ocean:

(a) Direct generation of long waves by atmospheric forcing (pressure and wind) on the sea surface;
(b) Generation of low frequency motions (e.g. storm surges) and subsequent transfer of energy into higher frequencies due to non-linearity, topographic scattering and non-stationarity of the resulting motions;
(c) Generation of high frequency gravity waves (wind waves and swell) and subsequent transfer of energy into larger scale, lower frequency motions due to non-linearity.

Long waves generated by the first two mechanisms are known as atmospherically induced or meteorological waves. The typical periods of these waves are from a few minutes to several hours, and typical length scales or wavelengths are from one to a few hundred kilometres. The first mechanism is the most important because it is this mechanism that is responsible for the generation of destructive seiche oscillations in particular bays and inlets of the World Ocean [12].

Meteorological waves can be produced by the passages of typhoons, hurricanes or strong cyclones. They have also been linked to frontal zones, atmospheric pressure jumps, squalls, gales, wind gusts and trains of atmospheric buoyancy waves. The most frequent sources of seiches in lakes are barometric fluctuations. They can also be produced by heavy rain and snow.

Infragravity waves are other types of long waves generated by the non-linear interaction of wind waves or swell. These waves have typical periods of 30 s to a maximum of 600 s and length scales from 100 m to 10 km. These waves are observed as sea level changes in the nearshore surf zone, where they have become known as surf beats. Infragravity waves have been found to be responsible for many phenomena in the coastal zone, including the formation of rip currents, wave set-up, sand bars, beach cusps and other regular forms of coastal topographies, as well as the transport of sediment materials. These waves can induce seiches in comparatively small scale semi-closed basins, such as ports and harbours, having natural periods of a few minutes and possibly posing a serious threat in the form of large amplitude wave responses.

A less frequent but often important generation mechanism in harbours is the tsunami or seismically formed ocean wave. Tsunami waves are the other main factors creating destructive seiche oscillations in bays, inlets and harbours. The 1946 Aleutian (magnitude M_w 8.6), 1952 Kamchatka (M_w 9.0), 1960 Chile (M_w 9.5) and 1964 Alaska (M_w 9.2) earthquakes induced strong seiche oscillations in bays, inlets and harbours throughout the Pacific Ocean.

In bays and gulfs, a dominant source of energy is the tide. Since there is quite a rich array of tidal frequencies, it is sometimes possible to determine the natural seiche frequency of a bay simply by observing the amplification of different tidal components. If the tidal period is close to that of the seiche period of the bay, resonance increases the tidal amplitude. Examples of this are the Bay of Fundy (Canada), where the tide can exceed 45 ft (15 m) and where a fundamental mode seiche period is about 13.3 h, close to the semidiurnal lunar tidal period of 12.4 h, and the Gulf of Mexico, where the period of the fundamental seiche mode is close to 24 h, with resulting amplification of the diurnal tide [12].

In smaller bays and harbours, the generation of seiches has been attributed to surf beats. This effect occurs because of the coupling of radiation stress, the thrust of the waves on the coast and rhythmic changes in the height of wind waves or swell. Since the swell can originate from distant storms, the existence of seiches is not necessarily closely correlated with local meteorological conditions.

In enclosed basins such as bays or lakes, seiches also occur by the shaking of the ground during earthquake excitation. Seismically induced water waves can be generated as a result of the motion of the boundaries enclosing the water body (basin) or through the tilting of the bed. Numerical studies conducted for Lake Union in Seattle, Washington, USA, showed that large sedimentary structures known for amplifying ground motions affect the distribution and magnitude of seiches [16–18]. It is also indicated that long, linear shorelines are the largest contributors to water wave heights. Depending on the shape of the lake, enhanced wave heights can also be a result of focusing. The study suggests that coastal structures can be vulnerable to the effects of resonant excitation initiated by earthquakes on water bodies far from tsunamigenic sources. The seiche amplification inside enclosed basins depends on the direction of the shaking (direction of the seismic wave), the shape of the basin and the reflection characteristics of the coastal boundaries.

3.3. BASIN AND HARBOUR OSCILLATIONS

Harbour oscillations (coastal seiches) are a specific type of seiche motion that occurs in partially enclosed basins (bays, fjords, inlets and harbours) that are connected through one or more openings to the sea. Resonant oscillations inside harbours, bays or any other semi-enclosed or closed basins are a problem that can have a direct effect on the management of harbours, shipping, cargo handling and coastal utilization of nuclear power plants. Surging due to long period oscillations in harbours is the result of a triple resonance of (i) external oscillations outside the harbour, (ii) natural oscillations within the harbour and (iii) natural oscillations of the vessels [12].

Specific spatial sea level and current variability occur during seiche oscillations. The intensity and direction of the currents vary significantly from place to place. The morphological characteristics in the basin, the orientation of the boundaries and the presence of structural elements (e.g. dams, dykes, piers and breakwaters) can create intense local vortexes. An example of this is the swirling current in Oarai harbour in Japan during the 2011 Great East Japan earthquake and tsunami.

The amplitude and persistence of the resulting seiches depend not only on the magnitude of the energy source but also on the energy losses within the water body. Such losses include dissipative effects resulting from friction on the sides or bottom of the basin. For semi-enclosed basins, energy can also be lost by the radiation of waves away from the mouth. In general, the rate of decay is greater for basins that are shallow or have narrow constrictions and complex topography.

The basic theory of seiche oscillations is similar to the theory of free and forced oscillations of mechanical, electrical and acoustical systems. The systems respond to external forcing by developing a restoring force that re-establishes equilibrium in the system. A pendulum is a typical example of such a system. Free oscillations occur at the natural frequency of the system if the system is disturbed beyond its equilibrium. Without additional forcing, these free oscillations retain the same frequencies, but their amplitudes decay exponentially with time as a result of friction until the system eventually comes to rest [12].

Analytical solutions for regular shaped, flat bottomed basins with vertical, solid, smooth and impermeable boundaries are available in the literature [19]. However, analytical methods cannot be as efficient at solving the problem of seiche oscillations in basins of irregular shape and bathymetry. In the case of the continental shelf, the incoming waves may be trapped, and seiches may be generated in the nearshore region. All these complexities may be addressed using numerical modelling.

Analytical solutions for regular (rectangular) shaped, flat bottomed basins with vertical, solid, smooth and impermeable boundaries are available in the literature [19, 20]. The periods of seiche oscillations (T_n) inside a closed, rectangular, flat bottomed basin are:

$$T = \frac{2}{\sqrt{gd}}\left[\left(\frac{n}{L}\right)^2 + \left(\frac{m}{B}\right)^2\right]^{(-1/2)}, n = 0, 1, 2, 3, \ldots, m = 0, 1, 2, 3, \ldots \tag{2}$$

where L and B are the short cut distances between solid boundaries of the rectangular basin, d is the water depth in the basin and n, m are integer numbers that represent each mode [21, 22].

If the basin is flat bottomed and semi-enclosed (i.e. one of the boundaries is an open boundary) then the periods of free oscillations — if the direction of the wave is perpendicular to the open boundary — become:

$$T = \frac{1}{n}\frac{4L}{\sqrt{gd}}, \ n = 0, 1, 2, 3, \ldots \tag{3}$$

The periods of free oscillations (T_n) inside a closed, circular, flat bottomed basin (when the boundaries are vertical, solid, smooth and impermeable) are [21]:

$$T_n = \frac{a_n 2\pi R}{\sqrt{gd}} \tag{4}$$

where R is the radius of the cylindrical section, d is the water depth, g is the gravitational acceleration and $a_1 = 1.64$, $a_2 = 0.896$, $a_3 = 0.618$.

Classically [22–25], several numerical computations can be performed to calculate the long wave amplitudes inside the basins in relation to the input waves with certain periods (see Section 8). When this test is repeated by inputting waves with different periods separately, the computed amplitudes can be compared, and the period causing the highest amplitude can be selected as one of the periods of free oscillations of the basin. This method entails many tests with many different input wave periods and requires a significant amount of time and effort for the determination of a single period of free oscillations. See

Section 6 for specifications of bathymetric and topographic data quality in the computation of seiches.

A short cut numerical method has been developed [20] as an alternative to the classical method, in which the periods of free oscillations (natural frequency) can be obtained directly as the output data of a single simulation in basins with irregular shapes and bathymetry. In the numerical application, the agitation inside the study basin is triggered by inputting an initial impulse (water level rise at the centre of the basin as a line crest, circular shape or any regular shape with a certain amplitude). The crest of the initial impulse may be either parallel to one of the coastal boundaries of the basin or as a vertical movement of a circular plane of water surface with a certain diameter. The numerical code solves the long wave equations and determines the time histories of water surface fluctuations at the numerical gauge locations of the study area. The spectrum curves of the computed sea level records at each location obtained by the fast Fourier transform technique show peaks at the periods of seiche oscillations, representing the periods of the seiche oscillations of the basin.

4. EFFECTS OF TSUNAMIS AT NUCLEAR POWER PLANT SITES

4.1. HAZARDOUS EFFECTS OF TSUNAMIS

The main effects of tsunamis at nuclear facilities relate to inundation: direct impact to the plant components by the strong forces resulting from the flow of water or its excess pressure. Figure 5 illustrates some of the effects of tsunamis on nuclear facilities discussed in this section.

Tsunami effects in the shallow zone and on land are directly related to the main tsunami hydrodynamic parameters. These are: (i) maximum positive amplitude, (ii) maximum current velocity, (iii) flow depth, (iv) hydrodynamic forces, (v) momentum flux, (vi) maximum negative amplitude, (vii) the arrival time of the first wave, (viii) the arrival time of maximum wave and (ix) the duration of the inundation and withdrawal. Different parameters related to the wave motion and morphology govern the magnitude of these parameters at the nuclear power plant site. The parameters must be determined under the conditions of design tsunamis (see Sections 5.3 and 9.1.1.3). The following subsections (4.1.1 to 4.1.8) provide the physical description of these parameters. Most of the phenomena related to them were observed in the 2004 and 2011 events.

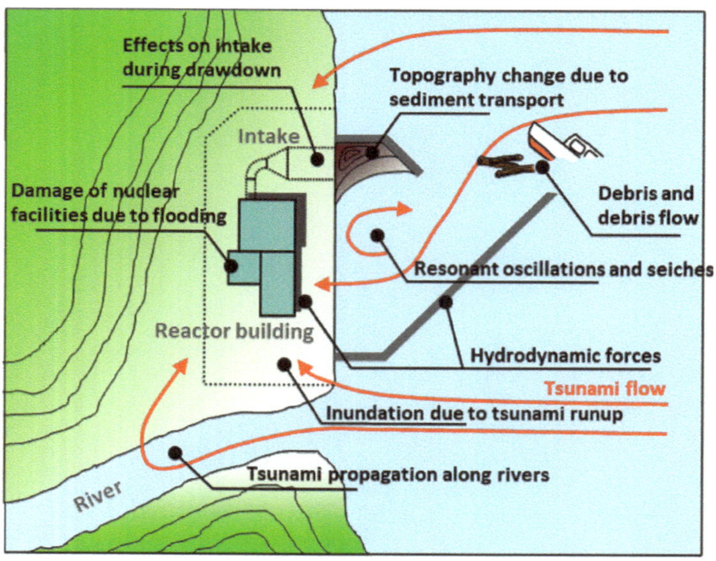

FIG. 5. Effects of tsunamis at nuclear facilities. Reproduced from Ref. [26] with permission courtesy of the Nuclear Regulation Authority of Japan.

4.1.1. Inundation due to tsunami runup

Tsunami flow in the shallow zone and on land depends on the amplitude of the incoming tsunami and its momentum (flow depth and current speed). The tsunami's energy may focus more on narrow bays, and its amplitude and momentum increase when it comes near shore. The boundary of the inundation zone (i.e. the land area flooded by the tsunami) is where the tsunami's energy dies out. The highest elevation from the still water level at the end of inundation (runup) depends on the tsunami's energy when it reaches the shallow water and on the topographic conditions of the coastal region.

After reaching the end point of the inundation, the water returns to the sea in a backwash motion. The water velocities during the backwash may exceed the velocities during the runup motion.

4.1.2. Tsunami propagation along rivers

When the tsunami enters a river mouth, the flooding waves can propagate for extensive distances along the riverbed. There have been many examples (such as the 2011 Great East Japan earthquake and tsunami) of tsunami penetration along rivers exceeding 10 km. Another pehnomenon was observed during the 1994 Mindoro tsunami in the Philippines, when, in an area where the vertical inundation heights did not exceed 3 m, the tsunami floated a 6000 tonne generating barge, broke its mooring lines and carried it 1.6 km inland along the Baruyan River [27].

4.1.3. Hydrodynamic forces

The depth and velocity of the flow are two major parameters that govern a tsunami's force. The drag force exerted by the tsunami flow is directly related to the cross-sectional area of the structure facing the flow and the square of the flow velocity. When the flow depth increases, the cross-sectional area of the structures and objects subjected to the flow increases. The flow velocity during the tsunami can increase more with the deeper flow depths. The mass density of the flowing material and shape factor (drag coefficient) of the structure are other parameters that govern the tsunami force.

Buoyancy is an upward force exerted by a fluid that opposes the weight of an immersed object. During tsunami inundation, a column of fluid surrounds the object. The pressure difference between the bottom and the top of the submerged object causes the buoyancy force that tends to move the object upwards. In a situation of fluid statics, the net upward buoyancy force is equal to the weight of the fluid displaced by the submerged object. The object floats when the buoyancy force exceeds its weight.

The shear force generated by the flow of water is exerted on the ground surface along the direction of flow. The flow velocity is a major parameter controlling the shear force. At the boundaries of stable, solid objects like buildings or other structures, the speed and direction of the flow pattern changes significantly, which also results in significant changes in the direction and magnitude of the shear force acting on the solid structures. Scour occurs when the shear force exceeds the magnitude of the force required to initiate the movement of the ground surface material. During tsunami inundation, boundary effects cause spatial changes in current velocity, and shear forces around structures thus sometimes cause unexpected depths of scouring.

4.1.4. Topography changes due to sediment transport

The maximum current velocity is important for erosion, deposition, debris and drag. Big shear forces on the seafloor cause deposition and erosion [1]. The eroded material is carried out and deposited where the energy of the flow becomes insufficient, such as near marine structures or at the nuclear power plant intake mouth. Unexpected deposition may cause operating problems in marine utilities and may even cause the partial or complete obstruction of the intake mouth, restricting or blocking the flow to the cooling system of the nuclear power plant. Scour is another problem that occurs around structures. The level of scouring is dependent on the flow speed and the material characteristics on the ground. The stability of the structure can be affected by the scour around the foundation.

4.1.5. Debris and debris flow

Debris material consists of a combination of bed and floating material (vegetation, sediment, human-made structures, etc.) in the sea and on land. The tsunami flow (forward or backwash) has sufficient energy to drag a huge mass of debris material. In this case, the mass density of the flowing fluid increases. By this effect, the impact and drag force of a tsunami increases tremendously, sometimes even more during backwash motion.

4.1.6. Resonant oscillations and seiches

Long wave oscillations can be generated inside enclosed or semi-enclosed basins and harbours (see Section 3). The periods of the long wave oscillations depend on the depth and shape of the basin. Amplification occurs when the period of the tsunami entering the basin fits the periods of free oscillations inside the basin. Potential resonance effects and consequent strong wave and seiche amplification near nuclear power plant sites should be investigated and computed by appropriate modelling (see Sections 7 and 8).

4.1.7. Effects on intake during drawdown

Continuous water intake for the cooling system with a sufficient volume rate at the nuclear power plant site is essential. The sea level drawdown that occurs during a tsunami or other critical event may cause a temporary or permanent lack of water input to the cooling system of a nuclear power plant.

4.1.8. Damage of nuclear facilities due to flooding

Most nuclear power plant facilities located inside the inundation zone of a tsunami may sustain damage to marine structures and their components (including coastal protection and water intake structures). Pumping stations, marine and land transportation networks, energy transmission structures, storage tanks, buildings and essential services (electricity transmission, water and other networks) may be affected by the impact of the tsunami's amplitude and momentum as well as by the accumulation of debris and the potential inundation of buildings and entire areas within the nuclear power plant site.

4.2. 2004 INDIAN OCEAN TSUNAMI

4.2.1. General

On 26 December 2004, a catastrophic tsunami struck the coastline of many countries facing the Indian Ocean, causing a significant loss of life and property. The tsunami waves were generated by a moment magnitude M_w 9.1 [28] subduction earthquake, the epicentre of which was located in the Andaman–Sumatra region of the Indian Ocean. Along the coast of mainland India, an estimated 8835 people lost their lives as a result of this calamity, and 86 people were reported missing [29]. Houses, roads, bridges and other infrastructure in the coastal region suffered great damage due to inundation by the tsunami waves. The 2004 tsunami also affected the nuclear power plant sites on the Indian coast.

India has three nuclear power plant sites located along the coast: Kalpakkam on the east coast, Kudankulam on the southern tip of the Indian peninsula and Tarapur on the west coast (see Fig. 6). The Kalpakkam site houses two 220 MW(e) pressurized heavy water reactors and a 500 MW(e) prototype fast breeder reactor

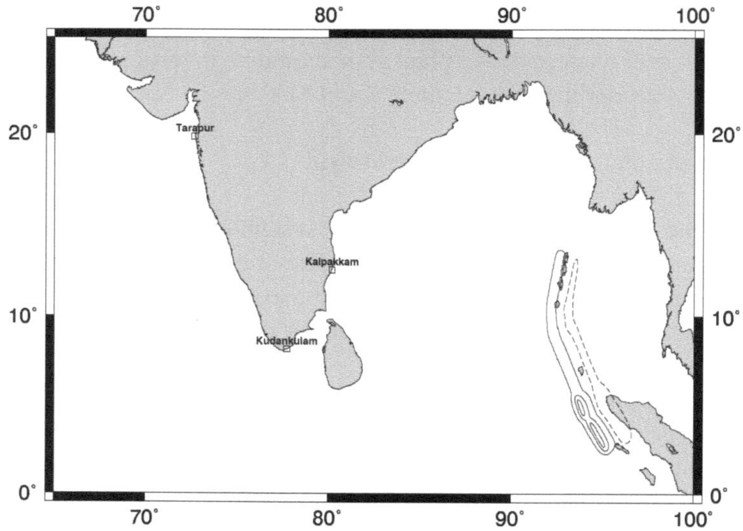

FIG. 6. Locations of the nuclear power plant and project sites along the Indian coast. A typical representation of the postulated initial water surface profile for the 2004 event is also shown (see Section 4.2.2 for proposed source models).

(PFBR) project. At Kudankulam, two 1000 MW(e) pressurized water reactors were under advanced stages of construction in 2004. The Tarapur site houses two 160 MW(e) boiling water reactors (BWRs) and two 540 MW(e) pressurized heavy water reactors. The tsunami runups were the highest in Kalpakkam and were negligible at Tarapur.

4.2.2. Earthquake and tsunami sequence

The 2004 Indian Ocean tsunami was generated by an undersea earthquake that occurred at 00:58:53 Coordinated Universal Time (UTC) on 26 December 2004. The earthquake's epicentre (3.316° N, 95.854° E) was located in the Indian Ocean just north of Simeulue Island, off the western coast of northern Sumatra, Indonesia, with a focal depth of 30 km. The estimated length of the rupture zone was approximately 1200 km with a width of about 150 km. The earthquake rupture's duration was several hundred seconds, and the maximum slip was approximately 30 m [30–33]. The United States Geological Survey [28] assigned M_w 9.1 to this earthquake. The mainshock was followed by more than 300 aftershocks of magnitude M_w greater than 5.0 within a month.

The 2004 tsunami was recorded around the world on tide gauges located in the Indian, Pacific and Atlantic Oceans. Its huge waves caused widespread destruction and the highest historically recorded number of casualties along the shores of several countries located around the Indian Ocean. The worldwide estimate of fatalities and missing persons due to this event is 297 200 [34]. In many locations along the shores of Sumatra, runups over 30 m were observed, with a maximum reported runup of about 50 m (see Fig. 7).

In addition to the tsunami, the earthquake caused other geophysical phenomena. Seiches were observed in India and the USA [34]. Subsidence and landslides were observed in Sumatra. A mud volcano near Baratang, Andaman Islands, became active on 28 December, and gas emissions were reported in Arakan, Myanmar [34].

After initiation, the tsunami waves took more than two hours to reach Indian shores. The wave arrivals were recorded at 09:05 Indian Standard Time (IST) at Chennai (near Kalpakkam) and Visakhapatnam on the east coast of India, at 09:57 at Tuticorin on the southern tip of the Indian peninsula and at 11:10 in Cochin. The wave heights along the east coast generally ranged from 3 m to 6 m, and a maximum wave height of 9.5 m was measured at Devanampattinam, Tamil Nadu (see Fig. 8).

Several tide gauges located along the coast of India and Sri Lanka recorded the tsunami waveforms. The records indicate that along the east coast of India (and the east coast of Sri Lanka), tsunami waves arrived shortly after high tide,

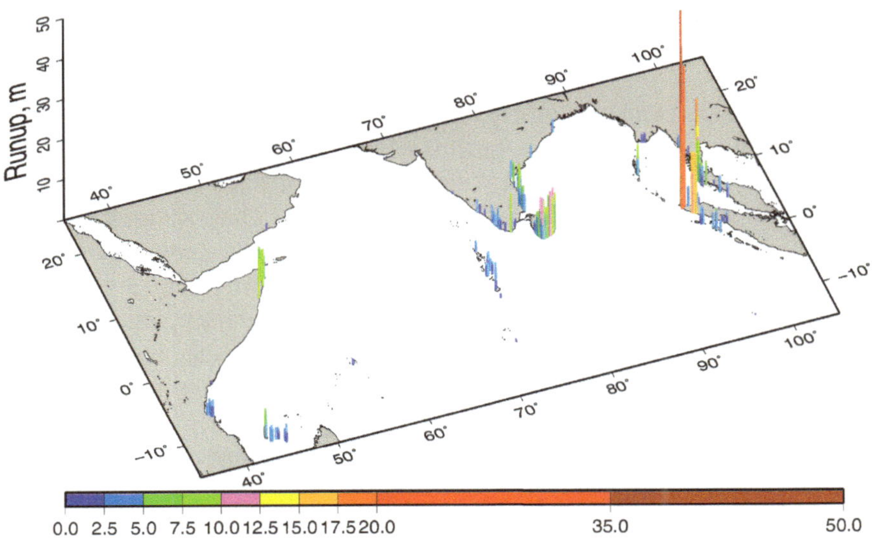

FIG. 7. Observed tsunami runups along the shores of the Indian Ocean based on the NCEI/WDS Tsunami Event Database (see Section A–1.1 in the Annex).

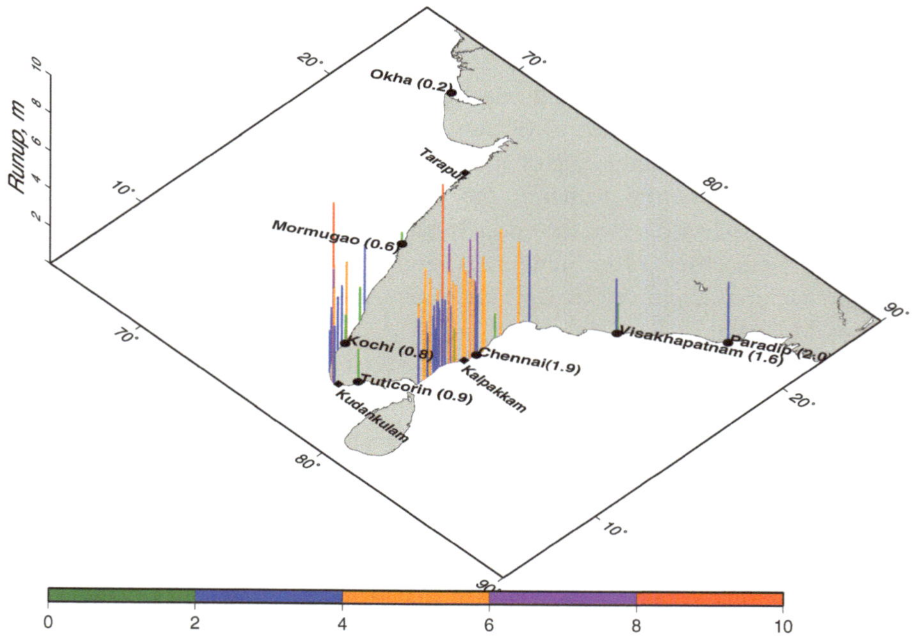

FIG. 8. Observed tsunami runups along the shores of India based on the NCEI/WDS Tsunami Event Database (see Section A–1.1). The nuclear power plant locations along the coast (squares) and tide gauge locations that recorded the event (circles) are also shown. The maximum water heights recorded by the tide gauges are given in brackets.

whereas in Tuticorin and Colombo, the waves arrived shortly after low tide [35]. All these gauges are located west of the earthquake zone, and the de-tided sea levels indicate a rise in sea level with the arrival of the tsunami, followed by a sharp decrease indicating the arrival of the trough. The locations of the tide gauges that recorded the event along the Indian coast are given in Fig. 8, and de-tided tsunami waveform data at these gauges are shown in Fig. 9.

Fourier analysis of tide gauge data indicates maximum amplitude at a period of 35–45 min, with another maximum at around 20 min. Also, along the Indian east coast, a broad peak was observed 1–2 h after the first wave arrival [35].

4.2.3. Effects and sequence of events at the nuclear power plants

4.2.3.1. Madras Atomic Power Station

Madras Atomic Power Station, located at Kalpakkam on the east coast of India, has two pressurized heavy water reactor units of 220 MW(e) capacity

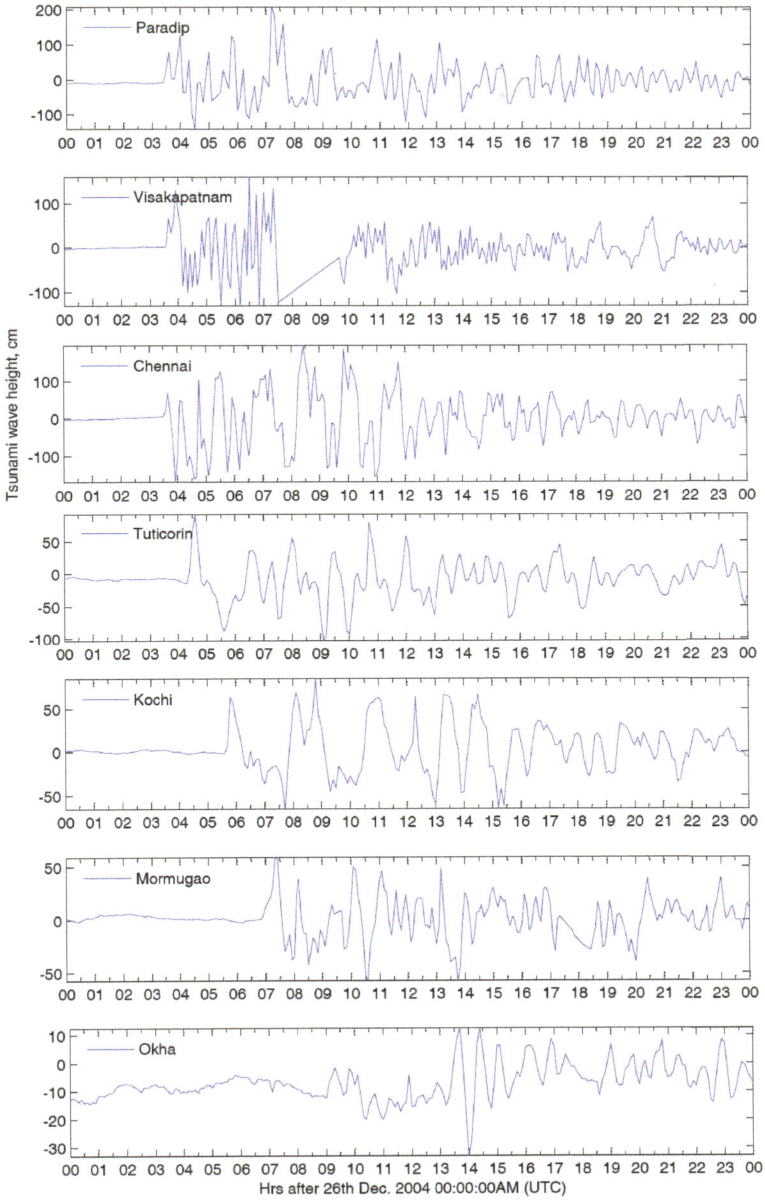

FIG. 9. De-tided tsunami waveform data at tide gauge stations based on National Institute of Oceanography (NIO) [36] and Nagarajan et al. [35]. Missing data have been joined with straight lines (e.g. Visakhapatnam tide gauge from 07:30 to 09:30).

each. Unit 1 was commissioned on 27 January 1984, and the second unit on 21 March 1986. Details of the nuclear power plant layout and the impact of the tsunami on the plant area are available in Refs [37–39].

The main structures in the plant area include the reactor building, the service building, the turbine building, the pump house and the intake structure, with the pump house being the closest structure to the shore (see Fig. 10). Sea water is used to provide cooling fluid to the turbine condenser and to cool the station process water, a closed loop system. The condenser cooling water pumps supply cooling water into the turbine condenser, and the process service water pumps supply water to the heat exchangers of the station process water system. From the intake structure located in the sea, the sea water is brought to the onshore pump house through an underground tunnel. The pumps are provided with screens to prevent the entry of debris, and their operating floor is about 2 m below the plant grade level.

The tsunami caused flooding in the seawater pump house of Madras Atomic Power Station at around 9:15 IST on 26 December 2004. The floodwaters entered the pump house through the underground intake tunnel connected to the onshore forebay of the pump house. Due to the high tsunami waves, the sea water level in the forebay increased to about 1.9 m above the operating floor of the pump house. During normal conditions, the sea water level is approximately 3 m below the operating floor.

Madras Atomic Power Station Unit 1 had been shut down for refurbishment at the time, but Unit 2 was in operation when the tsunami struck the shore of Kalpakkam. After the tsunami wave's arrival, the condenser cooling water pumps

1. Admin building
2. Warehouse
3. Canteen
4. DM water plant
5. DM water storage tanks
6. Sea water out fall
7. Sea water intake (not shown)
8. Approach jetty
9. Pump house
10. Turbine building
11. Reactor building
12. Service building
13. Stack
14. Upgrading plant
15. Indoor switch yard
16. Fresh water reservoir

FIG. 10. General plant layout of Madras Atomic Power Station Units 1 and 2 [37].

tripped as a result of the submergence of the local control switches in the sea water pump house. This affected the operation of the condenser cooling pumps of Unit 2. The operator tripped the turbine, followed by a controlled shutdown of the reactor. The reactor was cooled down and brought to a safe shutdown state as per normal design provisions and procedures, so all safety systems functioned as per design.

During the incident, one of the process service water pumps was in service providing sea water as cooling process water. Another process service water pump was put into service when the travelling water screen became clogged with debris transported by the tsunami.

The off-site power supply from the grid to the station continued to be available during and after the event. Though the off-site power remained available, emergency diesel generators (D/G-1 and D/G-5) were started and kept running as a precautionary measure. An emergency alert was declared at 10:25 IST on 26 December 2004 and was lifted at 21:43 IST on 27 December 2004 after the situation returned to normal.

A survey after the event indicated wave runups reaching about 4.5 m (relative to the mean sea level) in the site surroundings [40, 41]. The observed wave heights are given in Fig. 11. Apart from the water entry into the plant water pump house, other affected plant related structure, systems and components

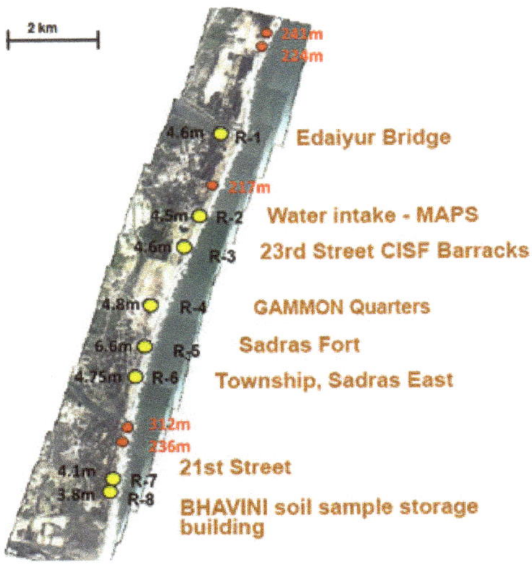

FIG. 11. Tsunami runup and inundation in the site area along the coast. Measured values are based on Singh and Kushwaha [40]. The points of measurement of wave runups are indicated by yellow circles, and the measured values are denoted by numbers in black text. The measurement locations of inundation distance are indicated by red circles, and the measured values are shown in red text.

include the dislocation of chlorination lines located on the jetty connecting the plant area to intake at sea and damage to a portion of the masonry wall near the outfall.

After the event, an on-site inspection was carried out. The event underwent a multitier review concerning operator response, emergency preparedness and the Atomic Energy Regulatory Board of India's requirements for the siting and design of nuclear facilities. After the rehabilitation of the pumps affected during the event and after obtaining the necessary regulatory clearance, the unit was restarted on 2 January 2005.

4.2.3.2. Prototype Fast Breeder Reactor, Kalpakkam

The 500 MW(e) PFBR site is located adjacent to Madras Atomic Power Station. The construction of the foundation raft of the PFBR building was in progress when the tsunami waves arrived, and the excavated foundation pit was flooded. Details of the effects of the tsunami on the plant area are available in Refs [42–44].

The excavation for the nuclear island was carried out in a plan area of about 150 m × 150 m, reaching 24 m depth from the evaluated plant grade level. Upon the completion of construction, the plant grade level would have been several metres above the natural ground level. The excavation created an 18 m deep pit below the mean sea level at a distance of about 200 m from the shoreline. Figure 12(a) depicts a view of the excavated pit just after the completion of the levelling course of plain cement concrete over base rock.

As per construction procedures, the 100 m × 100 m × 3.5 m thick foundation raft of PFBR is constructed in several layers or sectors. The construction of the

(a) (b)

FIG. 12. View of the excavated pit for the PFBR before and after the tsunami. Part (a) is a view of the excavated pit a few months before the tsunami [45]. Part (b) is a view of the excavated pit after the tsunami [46].

first layer of concrete was started on 15 December 2004. The second layer of raft concrete had been laid when the tsunami struck the site on 26 December 2004, filling the pit with 3.5 m × 10^5 m^3 of sea water, debris, sludge and sand.

The workers were safely evacuated before the tsunami waters inundated the foundation pit. Figure 12 shows the excavated pit of the PFBR before and after the tsunami.

The observed tsunami level at the PFBR site was about 2 m lower than the plant design basis flood level. The tsunami also did not exceed the flood levels estimated using tsunami levels prescribed in the regulatory document published in 1990, AERB/SG/S-11, Seismic Studies and Design Basis Ground Motion for Nuclear Power Plant Sites [47].

Following the removal of sand and debris brought in by the tsunami, massive rehabilitation work was carried out at the PFBR site. As the tsunami inundation took place during the time of concreting, it was postulated that the sea water could have changed the properties of the concrete. The rehabilitation work included an analysis of core samples from the seawater affected concrete concerning the raft pours for strength and durability, profiling the chloride, sulphate and cement content and the compressive strength. Chloride contamination was found to be concentrated only in the top 5 mm to 10 mm layer of concrete, and studies established that the chloride and sulphate content of the concrete were within acceptable limits once the top 10 mm layer of concrete was chipped off. Chloride migration studies were also conducted in order to work out options for the re-engineering of the raft. The cement content and the compressive strength of the concrete cores were found to be acceptable.

After reviewing the various technical issues and the various options for the restoration of the pit, it was decided to retain the layers of concrete constructed as part of the raft before the tsunami as a sub-base after chipping off the top layer of the concrete that was impregnated with chloride. Additional layers of highly impermeable plain concrete were poured over the existing sub-base pours, and freshwater proofing layers were laid over this. The entire raft was constructed afresh on top of it.

This reconstruction of the raft resulted in the raising of the level of the entire plant. After reconstruction, the finished floor level of the safety related buildings was about 9 m above mean sea level. This is about 3 m above design basis flood level and 5 m above the maximum water level rise observed during the 2004 tsunami.

Also, after the tsunami at Kalpakkam, shore protection and a tsunami embankment consisting of engineered boulders stacked at a height of 5.4 m above mean sea level were erected to reduce the momentum of sea waves during tsunami events.

4.2.3.3. Tsunami impact at Kalpakkam residential colony

The Kalpakkam township is situated approximately 4 km south of the PFBR site (see Fig. 11). Parts of the township that adjoin the coast suffered from the tsunami impact. Some people who were near the coast at the time were caught in the tsunami wave, leading to several fatalities in the area. Among them were five staff members from the Department of Atomic Energy [48]. In the township, the tsunami also caused the loss of power supply and the failure of communication lines [46].

4.2.3.4. Kudankulam Nuclear Power Plant

The Kudankulam site is located near the southern tip of the Indian peninsula. In 2004, the construction of two pressurized water reactors units of 1000 MW(e) each was in progress. During the tsunami event, there was no damage to the site. The inspection carried out after the tsunami event indicated that the maximum water level rise was 2 m above mean sea level, as determined from water level marks [49]. The site grade level is 7.5 m above the mean sea level and is thus much higher than the observed tsunami runup.

4.2.4. Lessons learned

Some of the issues identified as an outcome of the review of the Kalpakkam site after the 2004 tsunami were the following [38, 50, 51]:

- The need for a more rigorous treatment of tsunami hazards;
- A need for communication facilities to be augmented (e.g. in townships);
- The need for revision of emergency operating procedures for handling flooding incidents at all coastal sites;
- The need for relocation of some of the process equipment above the maximum flood levels at Madras Atomic Power Station;
- The need for the installation and implementation of a tsunami warning system.

4.3. THE 2011 GREAT EAST JAPAN EARTHQUAKE AND TSUNAMI

4.3.1. General

On 11 March 2011, a giant subduction earthquake (M_w 9.1) occurred off the Pacific coast of Tohoku, Japan, generating a devastating tsunami. This earthquake was the largest in Japanese history and the fourth largest in the world's instrumentally recorded history. The earthquake was officially named the 2011 Off the Pacific Coast of Tohoku Earthquake by the Japan Meteorological Agency (JMA), but it is often abbreviated to the Tōhoku (-oki) earthquake or the Great East Japan earthquake in scientific literature. The earthquake and tsunami disaster was officially named the Great Eastern Japan Earthquake Disaster by the Japanese Cabinet. Hereafter, the tsunami will be called the 2011 Great East Japan tsunami.

The earthquake and tsunami left approximately 15 900 people dead, 3 100 missing and 6 000 injured. Approximately 129 000 houses collapsed, 255 000 partially collapsed and 697 000 were partially damaged. Of these, some 7 600 houses were destroyed by the ground shaking, 19 000 were damaged by liquefaction, and the rest (nearly 98%) were affected by the tsunami. Approximately 500 000 people lost their homes and were displaced. The total economic loss is estimated at Y 16 900 billion (about US $200 billion), which represents approximately 20% of the annual national budget of Japan, according to the Central Disaster Management Council.

The four closest nuclear power plants to the Great East Japan earthquake source area are (see Fig. 13), from north to south, Onagawa nuclear power plant (Tohoku Electric Power Company), Fukushima Daiichi and Fukushima Daini (both Tokyo Electric Power Company) and Tokai Daini nuclear power plant (Japan Atomic Power Company). They all have BWR type reactors, and the number of reactors (units) is 14 in total. Of these, ten units were in operation during the 11 March Great East Japan earthquake.

4.3.2. Earthquake and tsunami sequence

The epicentre of the 2011 Great East Japan earthquake was 38.103° N, 142.860° E, and the origin time was 14:46:18 Japan Standard Time (JST) (UTC + 9 h), according to JMA. The earthquake was felt on most of the Japanese islands, reaching the highest seismic intensity degree of the JMA scale (7), which corresponds to XI or XII on the modified Mercalli scale or nearly 3000 gal (cm/s^2) of peak ground acceleration. The earthquake focal mechanism solution shows a thrust type fault movement on a gently dipping plane, indicating that it occurred along the boundary between the Pacific and the overriding Okhotsk plates. The Pacific plate subducts beneath northern Honshu along the Japan trench at a rate of about 8 cm per year. This subduction causes westward movement and east–west compression of northern Honshu, as observed before 2011 by the land based Global Positioning System (GPS) network [53].

The 2011 Great East Japan earthquake was preceded by foreshocks. Seismic activity, including four earthquakes with $M_w \sim 5$, occurred in February 2011. On

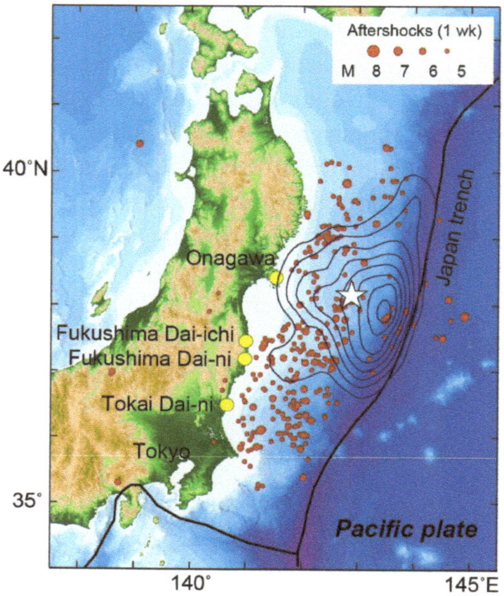

FIG. 13. Epicentres of the mainshock and aftershocks within one week of the Great East Japan earthquake. The white star is the epicentre of the mainshock, and the red circles are aftershocks that occurred within one week, according to JMA. The black contours indicate the slip amount (contour interval of 4 m) on the source fault, estimated from the tsunami waveforms [52]. The locations of the four nuclear power plants referred to in the text are shown as yellow circles.

9 March, two days before the mainshock, an M_w 7.3 earthquake occurred about 45 km north-east of the mainshock epicentre. This earthquake, now recognized as a foreshock, had a mechanism similar to that of the mainshock, with east–west compression, and produced a minor tsunami with a maximum height of 0.6 m at Ofunato. Intense seismic activity, including an M_w 6.5 event, continued until the mainshock on 11 March.

The Great East Japan earthquake produced numerous aftershocks. Two large aftershocks with $M_w > 7$ occurred within 40 min of the mainshock. Within two months (up to 10 May), three $M_w \geq 7$ earthquakes (including the above two), 60 $M_w \geq 6$ earthquakes and 610 $M_w \geq 5.0$ earthquakes took place over a source area 500 km long and 200 km wide. The large 7 April aftershock (M_w 7.1) was deeper than the plate interface and is considered to have occurred within the subducting Pacific plate.

The Great East Japan earthquake also triggered earthquakes in central to eastern Japan. Three $M_w > 6$ earthquakes occurred at large distances (more than a few hundred kilometres away) within seven days of the mainshock. These are also interpreted as triggered events. In Fukushima Prefecture, an M_w 6.6 normal fault earthquake, reflecting east–west extensive stress, occurred on 11 April. It was followed by many aftershocks with similar focal mechanisms.

The slip on the fault of the 11 March mainshock was estimated to be as large as 50 m from various geophysical data such as seismic waves [54], land GPS [53] and marine geodetic measurements [55] or tsunami waveforms [52]. The fault slip caused a significant seafloor displacement, which was the mechanism of the devastating tsunami that followed.

The 2011 Great East Japan tsunami was an unprecedented tsunami disaster in the history of Japan. It was instrumentally recorded by various types of gauges, including coastal tide and wave gauges, offshore GPS buoys, cabled ocean bottom pressure gauges and Deep-ocean Assessment and Reporting of Tsunamis (DART) buoys. The tsunami reached the coast of Iwate, Fukushima and Ibaraki Prefectures 30 min after the earthquake. The tsunami inundated coastal areas from Hokkaido Prefecture in the north to Chiba Prefecture in the south.

The coastal tsunami heights, both inundation heights and runups, were measured by the Joint Survey Group, which consisted of approximately 300 researchers from more than 60 organizations [56] (see Fig. 14). The total number of measurement points exceeded 5900. The density of measurements far exceeds those available for any previous tsunamis, providing invaluable data for future research. The disaster scale is described by the maximum tsunami heights and

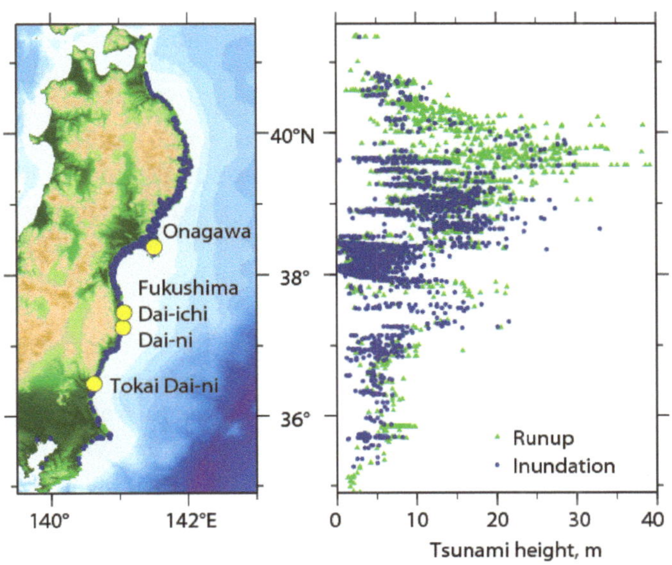

FIG. 14. Locations of the four nuclear power plants around the source areas (left) and distribution of measured tsunami heights (right) [56].

the length of the coast affected by large tsunami waves. The former reached more than 35 m at the coast of Iwate Prefecture, which is similar to the tsunami runup of the 1896 Sanriku tsunami. Runups greater than 10 m were observed across more than 530 km of coastline.

Figure 15 shows examples of tsunami waveforms recorded at an offshore bottom pressure gauge (located about 70 km from the coast at a depth of 1600 m), a GPS wave gauge (located around 10 km from the coast at a depth of 200 m) and four coastal tide gauges. The offshore bottom pressure gauge shows that the sea level started to rise while the seafloor was shaking with seismic waves, followed by a large impulsive peak at 12 min. Similar two stage tsunami waveforms with large amplitude and delayed start (15 min) were recorded at the GPS wave gauge. The Kamaishi and Ayukawa tide gauges recorded the tsunami's arrival at 25 min, but the large tsunami destroyed the instrument before it could record the largest peak. These waveforms, as well as those on DART buoys, were used to estimate the tsunami source [52, 57, 58].

4.3.3. Effects and sequence of events at nuclear power plants

4.3.3.1. Fukushima Daiichi nuclear power plant

The Fukushima Daiichi nuclear power plant (Fig. 16) is located in Fukushima Prefecture and consists of six BWR units. Unit 1 was commissioned in March 1971. The total generating capacity of all units is 4696 MW(e).

At the time of the earthquake on 11 March 2011, Units 1, 2 and 3 were under operation, and Units 4, 5 and 6 were under periodic inspection. The earthquake disrupted all of its six external power supply sources, which resulted in the startup of its emergency diesel power generators (D/G). However, the ensuing tsunami strike caused the plant to undergo a 'station blackout' (i.e. a total loss of alternating current (AC) power) by submerging or flooding the emergency D/Gs, AC power supply equipment, metal-clad switchgear, power centre and other electrical distribution equipment, rendering it useless.

The tide gauge in the harbour was damaged by the tsunami, so the water level could not be recorded. The ultrasonic wave gauge approximately 1.5 km offshore from Fukushima Daiichi nuclear power plant was damaged by the second tsunami wave and collected data only until 15:35 JST. The water level began rising at around 15:15 JST and, after a gradual rise, peaked at about 15:27 JST. Next, the water level briefly dropped before suddenly jumping up at 15:33 JST and then exceeding the measurement limit of OP + 7.5 m (see Fig. 17). The abbreviation OP (for Onahama Peil) is the reference sea level at the port of Onahama, which is 0.727 m below the mean sea level of Japan, denoted Tokyo Peil (TP).

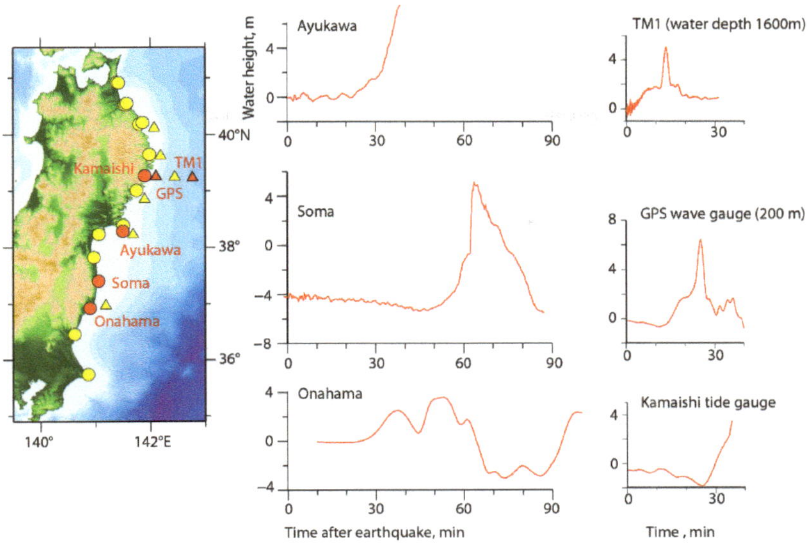

FIG. 15. Tsunami waveforms for the 2011 Great East Japan tsunami recorded at the stations shown by red symbols in the image on the left. They are the offshore bottom pressure gauge (TM-1) and GPS gauge (triangles) and four tide gauges (circles). Other stations shown by yellow symbols also recorded tsunami waveforms, but they are not shown here.

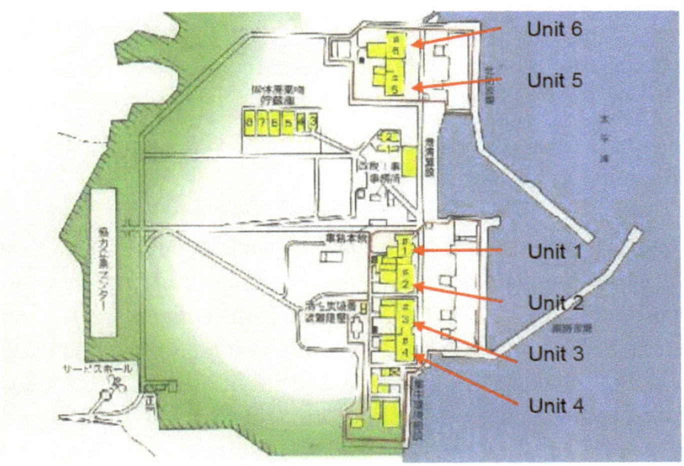

FIG. 16. General layout of Fukushima Daiichi nuclear power plant. Reproduced from Ref. [59] with permission courtesy of the Nuclear Emergency Response Headquarters of Japan.

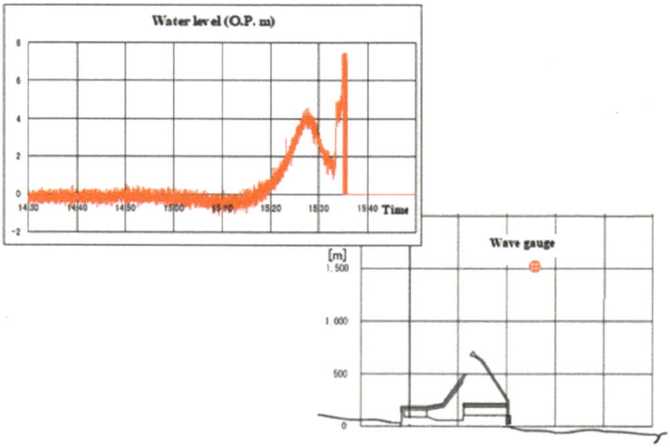

FIG. 17. Water level observed at Fukushima Daiichi nuclear power plant. The measurement upper limit was OP + 7.5 m. Reproduced from Ref. [60] with permission courtesy of the Nuclear Regulation Authority of Japan.

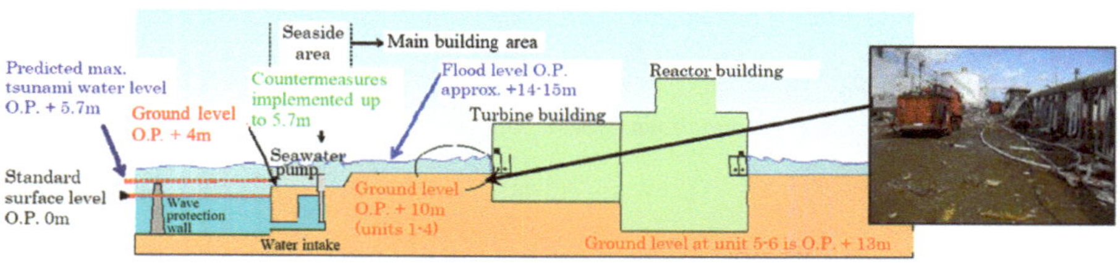

FIG. 18. Schematic cross-section of Fukushima Daiichi nuclear power plant. Reproduced from Ref. [61] with permission courtesy of the Nuclear Regulation Authority of Japan.

(a) Flooding due to tsunami runup

The tsunami at the Fukushima Daiichi nuclear power plant inundated the seaside area (ground level OP + 4 m) and the main building area (ground level OP + 10 m around Units 1 to 4; OP + 13 m around Units 5 and 6), and the entire seaside area and main building area were flooded (see Fig. 18). The inundation height was approximately OP + 11.5 m to 15.5 m around Units 1 to 4, and the flow depth was approximately 1.5 m to 5.5 m (see Fig. 19). In Fig. 19 'Inundation depth' means flow depth. The inundation height around Units 5 and 6 was approximately OP + 13 m to 14.5 m, and the flow depth was approximately 1.5 m or less, which was relatively shallow compared with the area around Units 1 to 4.

Pictures of the tsunami near the building on the south side of Unit 4 show a tank approximately 5.5 m in height installed at an elevation of OP + 10 m being

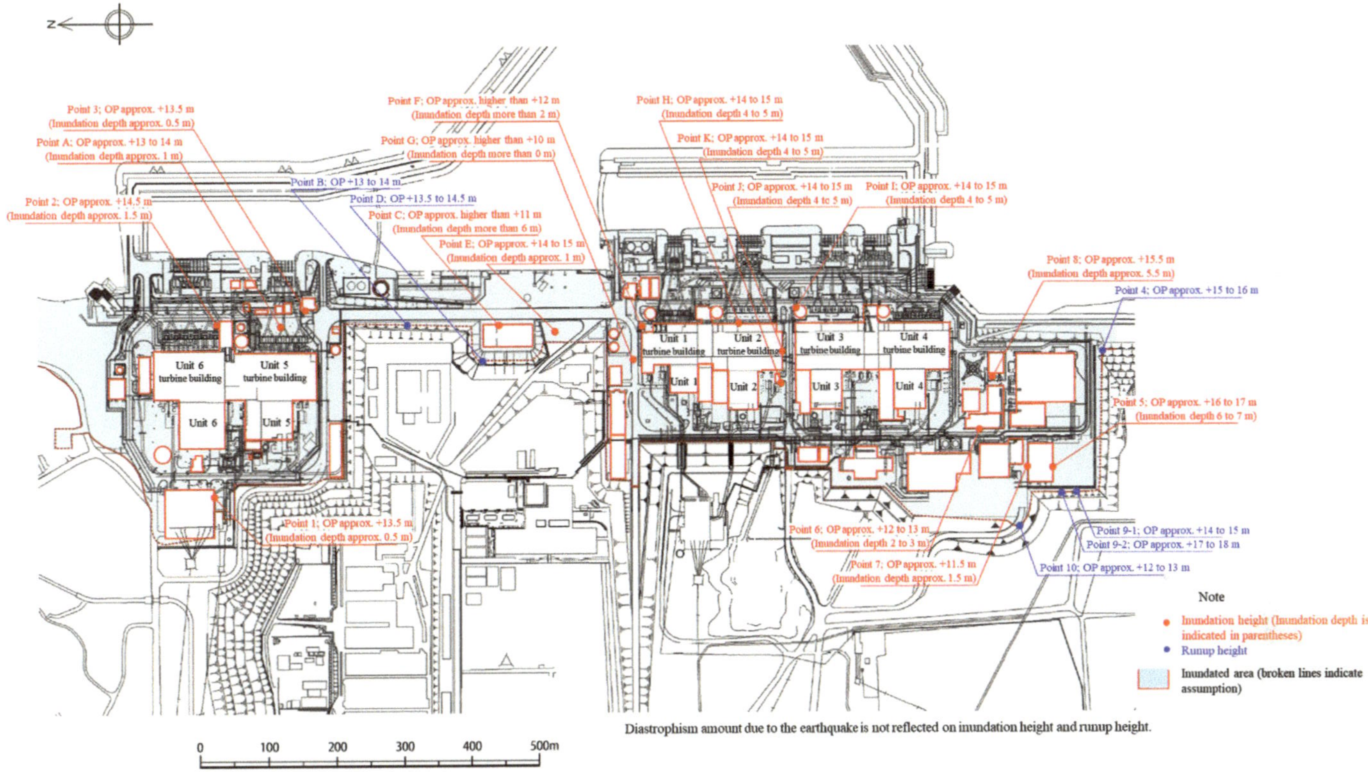

FIG. 19. Inundation height (flow depth) and runup at Fukushima Daiichi nuclear power plant. Reproduced from Ref. [62] with permission courtesy of the Nuclear Emergency Response Headquarters of Japan.

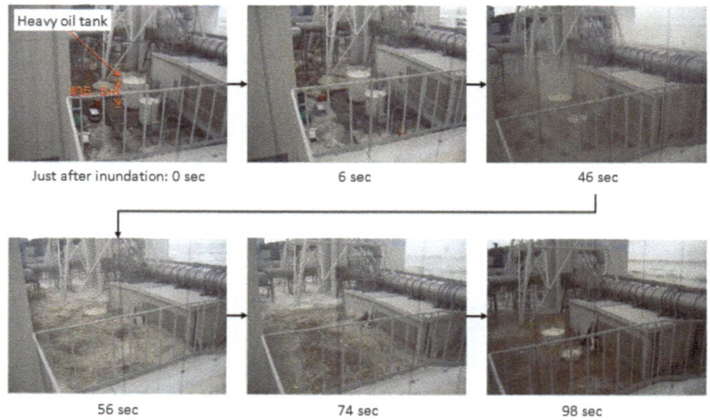

FIG. 20. Inundation near Unit 4 at Point 8 of Fig. 19. Reproduced from Ref. [63] with permission courtesy of Tokyo Electric Power Company.

submerged by the tsunami. The flow depth in the vicinity of the building was more than 5 m above the ground (see Fig. 20).

(b) Hydrodynamic forces

Figure 21 shows that there are two smaller tanks (heavy oil tanks) and a large tank (suppression pool tank). One of the smaller tanks was moved by the tsunami, and the large one was damaged.

Several pieces of the concrete crown wall of the breakwater were overturned and displaced toward the land side during the tsunami (see Fig. 22). There was minor damage to the rubble mound armour layer.

Significant damage to the structure of the main building, such as the outer wall and pillars, etc., was not found. It was also confirmed that tsunami damage mainly occurred on the east side (the seaside) of Units 1 to 4 of the turbine buildings, parts of doors and shutters, etc.

However, some pumps, as well as ancillary equipment, were damaged by the impact of a collapsed crane used for facility inspection and floating objects entrained in the tsunami.

Before the tsunami, two pumps had been removed for inspection from their originally installed locations. Those pumps, under inspection, were displaced by the tsunami. After the tsunami, the other pumps that were not under inspection remained at their originally installed locations. However, the tsunami caused them to lose function.

FIG. 21. Situation of damaged structures. Reproduced courtesy of Tokyo Electric Power Company [64].

(c) Morphological changes

The morphological changes were investigated using single beam and narrow multibeam sounding systems.

Erosion was found around the harbour entrance; the maximum amount of erosion was approximately 3.5 m. Deposition was found throughout the harbour and the intake canal (see Fig. 23).

(d) Debris and debris flow

A large amount of floating debris remained in the vicinity of the Fukushima Daiichi nuclear power plant (see Figs 24 and 25). This included the heavy oil tank (with a diameter of 11.7 m, a height of 9.2 m and a weight of 32 tonnes) that had been installed on the seaward side (ground level: OP + 4 m) and was displaced by the tsunami to the road on the north side of the reactor and turbine buildings at Unit 1 (ground level: OP + 10 m). Many cars were also washed away.

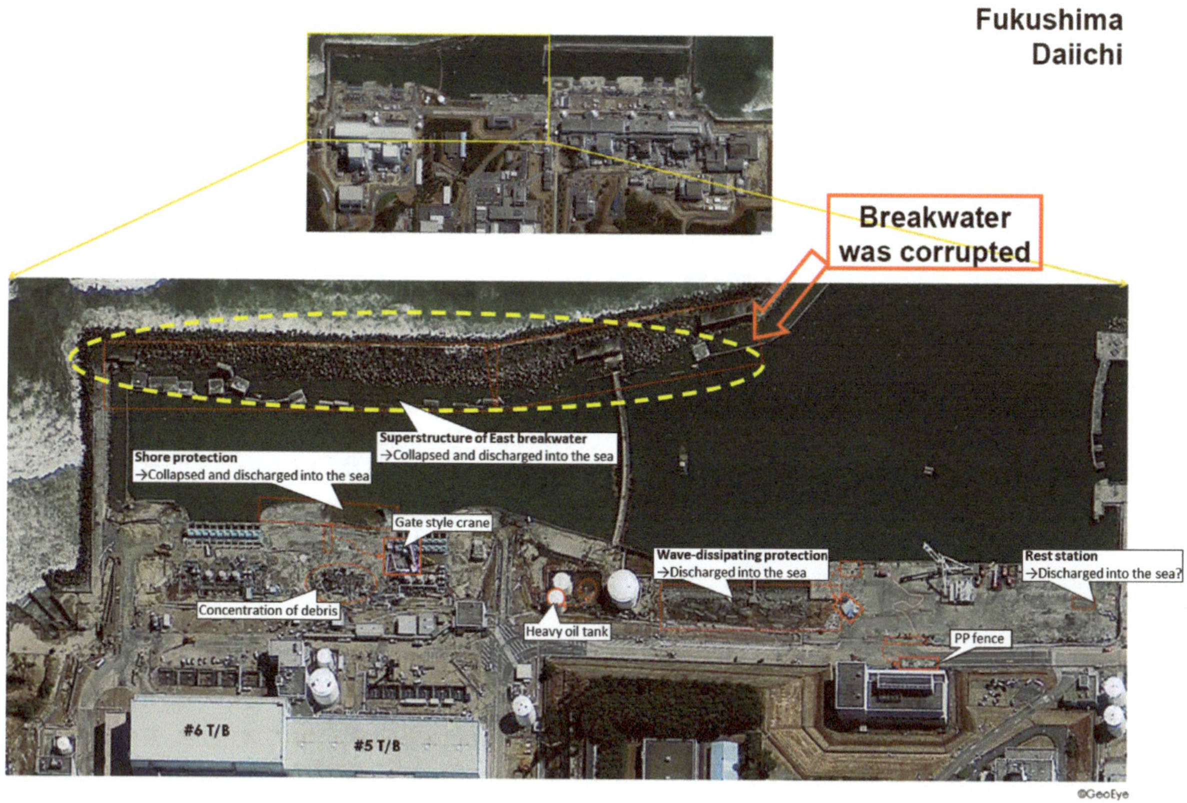

FIG. 22. *Location of breakwater and tanks. Reproduced courtesy of Tokyo Electric Power Company [64].*

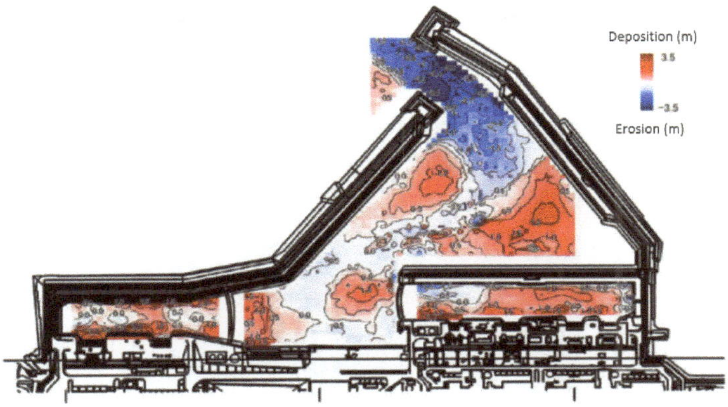

FIG. 23. Morphological change at Fukushima Daiichi nuclear power plant. Reproduced courtesy of Nuclear Regulation Authority of Japan [60].

FIG. 24. Debris at Fukushima Daiichi nuclear power plant. Reproduced courtesy of Tokyo Electric Power Company [64].

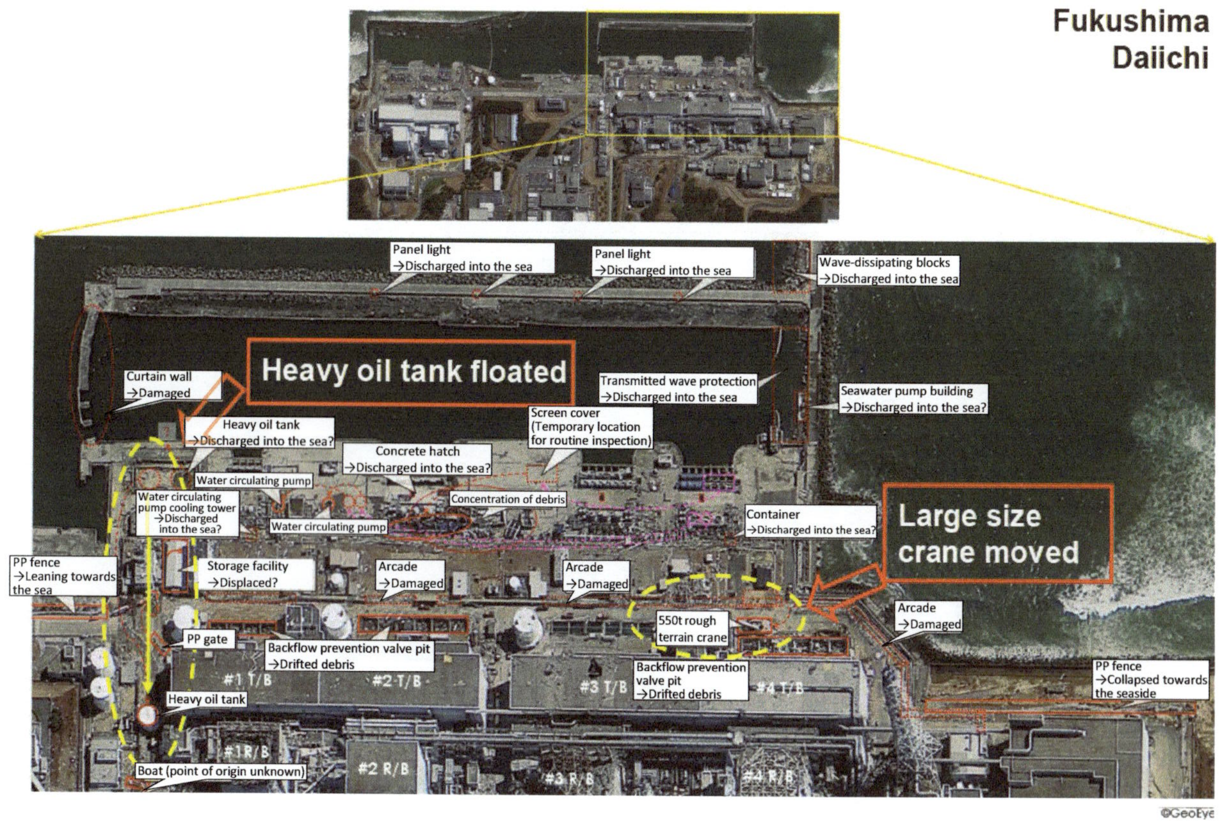

FIG. 25. *Location of debris at Fukushima Daiichi nuclear power plant. Reproduced courtesy of Tokyo Electric Power Company [64].*

(e) Inundation damage of nuclear facilities

Some parts of the openings on the ground level of the main building (building entrance, equipment hatch and exhaust port) and the openings of the trenches and ducts buried underground at the site (penetration slots for cables and pipes) were assessed to have provided flooding routes into the buildings (see Figs 26 and 27). It is estimated that, inside the buildings, a large area of the basement floors was flooded through passageways and stair shafts.

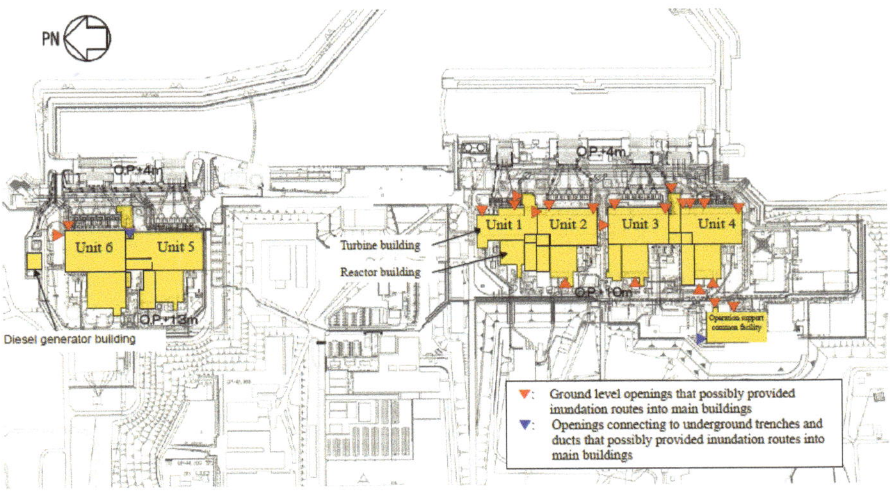

FIG. 26. *Locations of openings that possibly provided inundation routes into the main buildings at Fukushima Daiichi nuclear power plant. Reproduced courtesy of Nuclear Emergency Response Headquarters of Japan [62].*

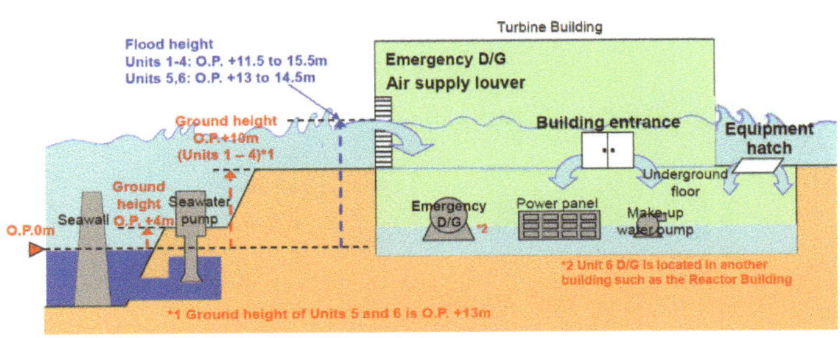

FIG. 27. *Flood pathways into the major buildings. Reproduced courtesy of Tokyo Electric Power Company [63].*

4.3.3.2. *Fukushima Daini nuclear power plant*

Fukushima Daini nuclear power plant is located near the towns of Tomioka and Naraha of Futaba County in Fukushima Prefecture. The commercial operation of Unit 1 started in April 1982, and there were four BWR units at the time of the event (see Fig. 28), the total generating capacity of which was 4400 MW(e).

All of the nuclear reactors, Units 1 to 4, at the Fukushima Daini nuclear power plant were in operation. The reactors from Units 1 to 4 at Fukushima Daini nuclear power plant tripped at 14:48 JST on 11 March. A total of four external power supply lines were connected to this nuclear power plant. One line was under maintenance and was not functional; two lines ceased to function due to the earthquake (one immediately and the other after one hour), which resulted in the electric supply being provided by a single line (restoration work on the third line was completed at 13:38 JST on 12 March, so two lines became available after that). The nuclear power plant was hit by the tsunami at around 15:34 JST on 11 March, resulting in a loss of residual heat removal systems for all units except Unit 3.

In Unit 3, one of the two residual heat removal systems survived, and cooling continued after the tsunami's arrival. Therefore, the reactor reached cold shutdown status at 12:15 JST on 12 March. However, in Units 1, 2 and 4, the temperature of the suppression pool water exceeded 100°C because not all of the heat could be removed. The cooling of the suppression pool started with the operation of the residual heat removal system by connecting a temporary cable from the functioning distribution board. As a result, each reactor reached cold shutdown status on 14 or 15 March.

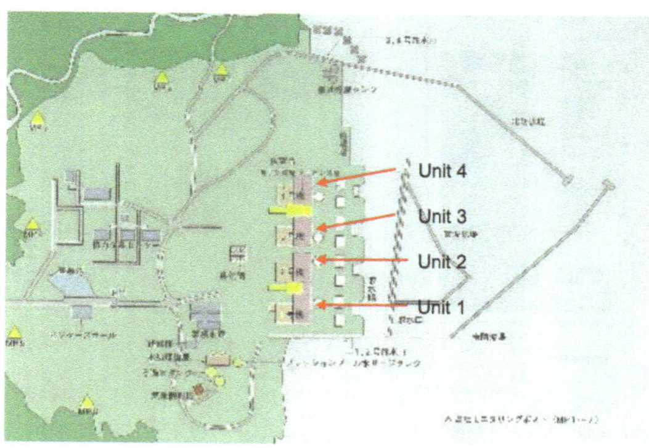

FIG. 28. General layout of Fukushima Daini nuclear power plant. Reproduced courtesy of Nuclear Emergency Response Headquarters of Japan [59].

A tide gauge was installed in the harbour, but, as it had been damaged by the tsunami, the water level could not be recorded at Fukushima Daini nuclear power plant.

(a) Flooding due to the tsunami runup

The tsunami at the Fukushima Daini nuclear power plant inundated the entire seaside area (OP + 4 m) in front of Units 1 to 4 and partially inundated the main building area (OP + 12 m) around Units 1, 2 and 3 (see Figs 29–31). The tsunami inundated from the narrow space between the south side of Unit 1 and the slope in the main building area and reached the main building areas of Units 1, 2 and 3. The inundation height was approximately OP + 12 m to 14.5 m around Units 1 to 3, and the flow depth was approximately 2.5 m or less, which was relatively shallow compared with Fukushima Daiichi nuclear power plant. In Fig. 29, 'flooded depth' means flow depth. The main building area of Unit 4 was not inundated.

(b) Hydrodynamic forces

In all units, the doors of the heat exchanger buildings and other smaller structures located in the seaside area (OP + 4.0 m) were damaged by the tsunami (see Fig. 32).

Only the doors, louvers and hatches along the southern half of the main buildings of Unit 1, located in the main building area (OP + 12 m), were damaged by the tsunami. No damage was confirmed at the main building's outer walls and pillars.

(c) Morphological changes

Erosion was found around the harbour entrance; the maximum amount of erosion was approximately 3 m. Deposition was found in the centre of the harbour and the front of the shallow draft quay (see Fig. 33).

(d) Debris and debris flow

Temporary equipment for inspection, the steal stop logs, etc., were displaced by the tsunami. At the Fukushima Daini nuclear power plant, none of the major equipment and structures was found to have drifted to the main building area (ground level: OP + 12 m) as a result of the tsunami.

Figure 34 shows the debris near the heat exchanger building of Unit 3.

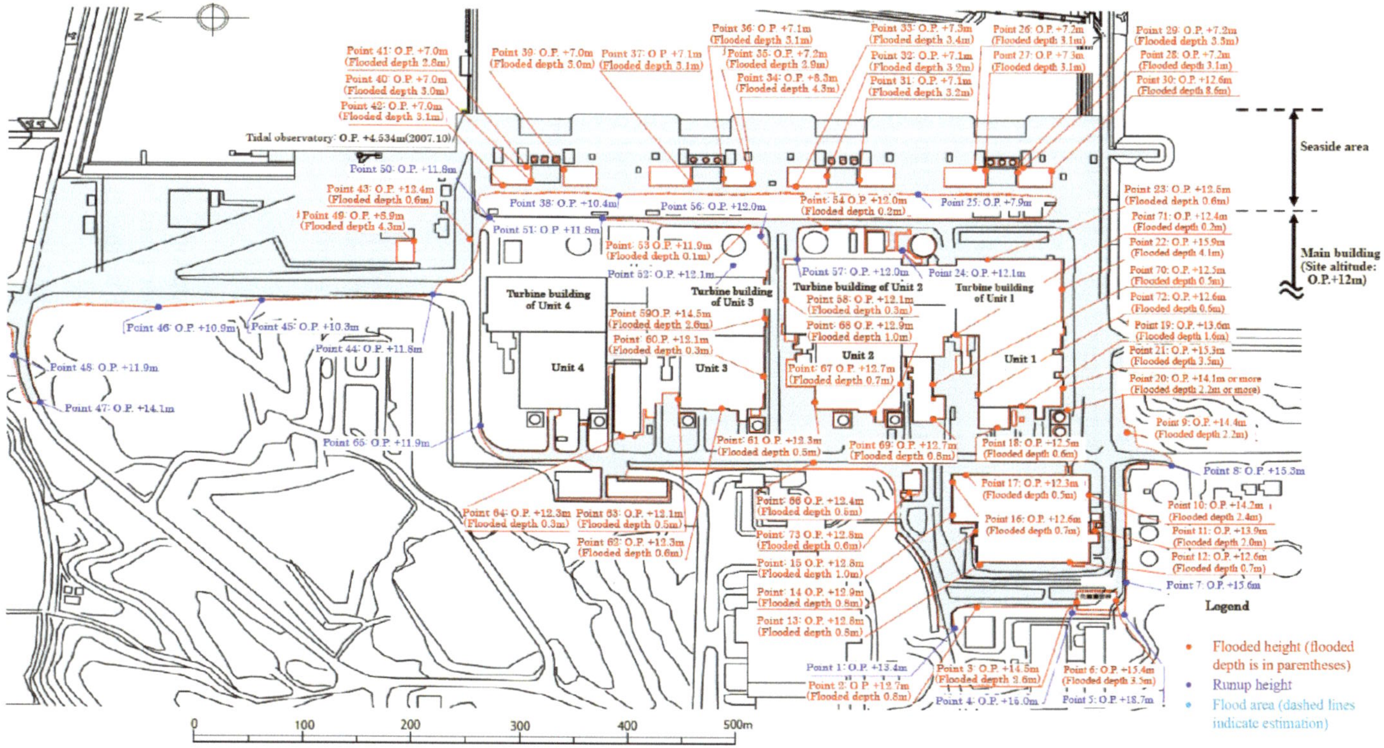

FIG. 29. Inundation height (flow depth) and runup at Fukushima Daini nuclear power plant. Reproduced courtesy of Nuclear Emergency Response Headquarters of Japan [62].

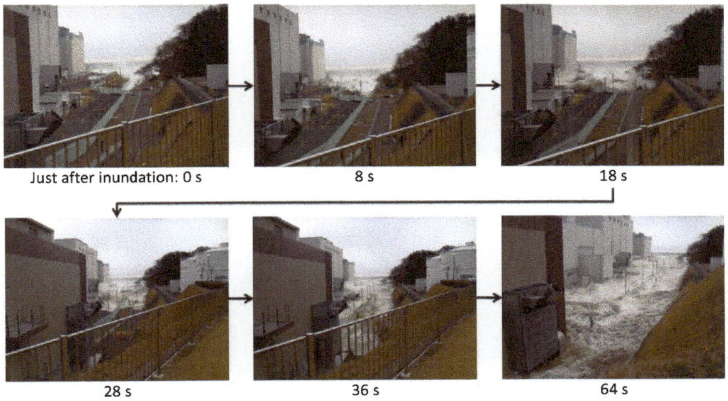

FIG. 30. Tsunami runup along the slope of Fukushima Daini nuclear power plant. Reproduced courtesy of Nuclear Regulation Authority of Japan [60].

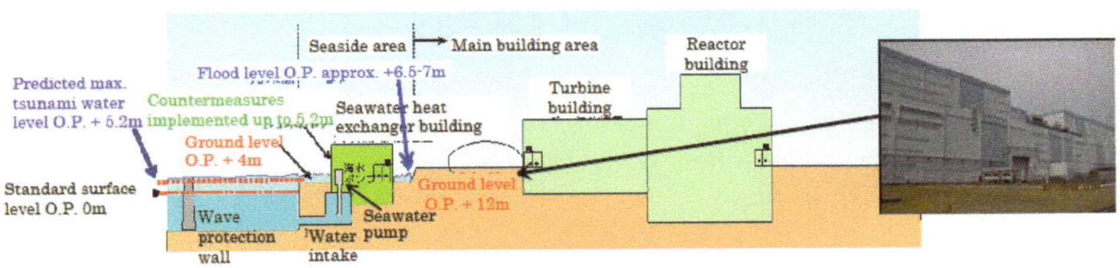

FIG. 31. Schematic cross-section of Fukushima Daini nuclear power plant. Reproduced courtesy of Nuclear Regulation Authority of Japan [61].

FIG. 32. Damaged door of the heat exchanger buildings. Reproduced courtesy of Tokyo Electric Power Company [64].

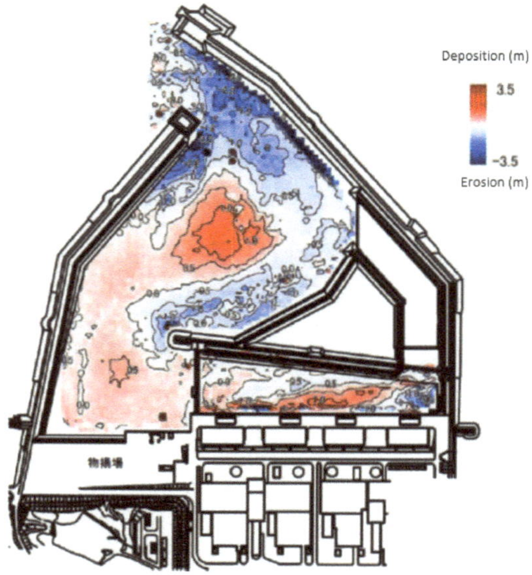

FIG. 33. Morphological change at Fukushima Daini nuclear power plant. Reproduced courtesy of Nuclear Regulation Authority of Japan [60].

FIG. 34. Debris near the heat exchanger building of Fukushima Daini nuclear power plant, Unit 3. Reproduced courtesy of Tokyo Electric Power Company [64].

(e) Inundation damage of nuclear facilities

Emergency seawater system pumps were installed in the heat exchanger buildings at Fukushima Daini nuclear power plant. The area around the heat exchanger building was flooded to a level of about 3 m by the tsunami. Although the frameworks of these buildings were not damaged, the doors and other ground level openings were damaged, and all heat exchanger buildings were flooded (see Figs 35 and 36). As a result, almost all power panels and pump motors ceased to function because of water damage.

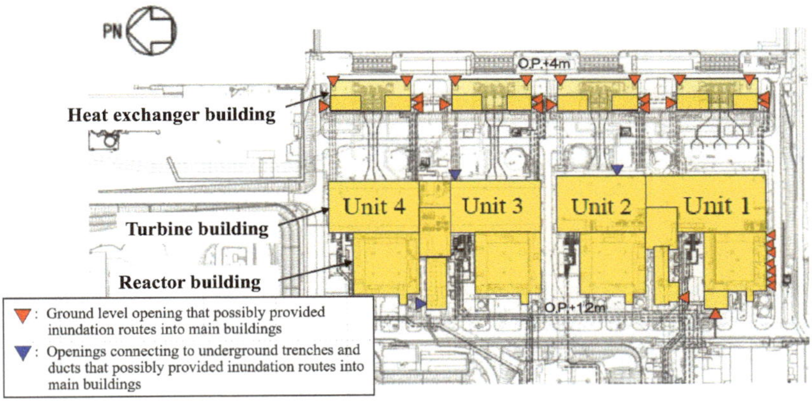

FIG. 35. Locations of openings that possibly provided inundation routes into the buildings at Fukushima Daini nuclear power plant. Reproduced courtesy of Nuclear Regulation Authority of Japan [60].

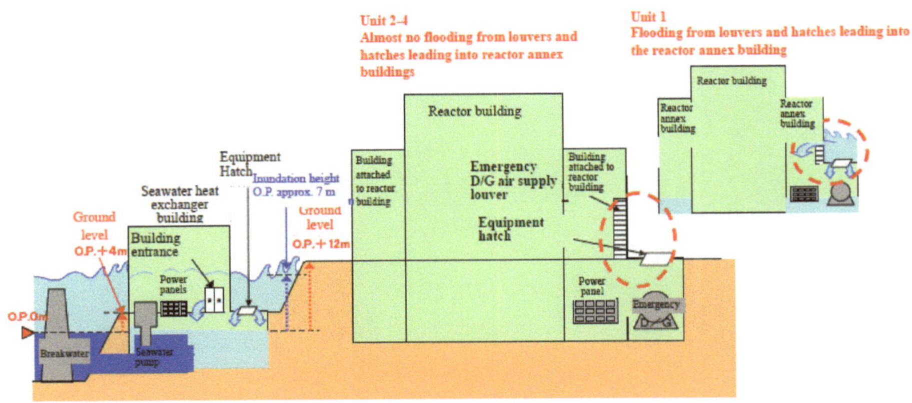

FIG. 36. Flood pathways into the major buildings. Reproduced courtesy of Tokyo Electric Power Company [63].

Three emergency D/Gs were installed for each unit. At Unit 1, water flooded the reactor annex building from the ground level openings, and all three emergency D/G main units stopped working. At Unit 2, none of the D/G main units were damaged but they all ceased to function because the power panels or the pump motors of the emergency D/G seawater cooling systems were damaged by water. For Units 3 and 4, none of the emergency D/G main units were damaged by flooding, and some emergency D/G seawater cooling systems were also available. In total, nine subsystems of emergency D/Gs ceased to function, and three continued to function. However, there was no need to use the remaining emergency D/Gs, as off-site power continued to be available at Fukushima Daini nuclear power plant.

4.3.3.3. Onagawa nuclear power plant

The Onagawa nuclear power plant is located near Onagawa Town, in Oshika County and near Ishinomaki City, in Miyagi Prefecture. The commercial operation of Unit 1 started in June 1984, and there are three BWRs now, the total generating capacity of which is 2174 MW(e) (see Fig. 37).

Units 1 and 3 were in operation at the time the earthquake occurred on 11 March, and Unit 2 was under reactor startup operation. The off-site power supply from four out of the five lines was interrupted as a result of the earthquake, but it was maintained through the continued operation of one power line.

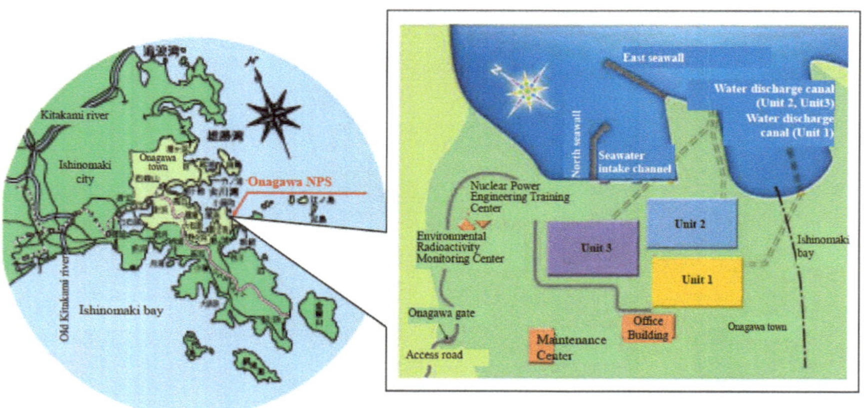

FIG. 37. General layout of the Onagawa nuclear power plant. Reproduced courtesy of Nuclear Emergency Response Headquarters of Japan [62].

The reactor at Unit 1 tripped at 14:46 JST because of high seismic acceleration, and the emergency D/Gs (A) and (B) started. The cooling and depressurization operations of the nuclear reactor were performed successfully, and the reactor reached a state of cold shutdown with a reactor coolant temperature of less than 100°C at 00:57 JST on 12 March.

Since the reactor was in startup operation, Unit 2 shifted promptly to cold shutdown because the reactor tripped automatically at 14:46 JST as a result of the significant seismic acceleration. Subsequently, the emergency D/G (B) was tripped because the reactor's auxiliary component cooling water system B pump and other cooling system pumps were inundated as a result of the tsunami and thus lost function. However, since component cooling water system A pump was intact, there was no influence on the reactor's cooling function.

The reactor at Unit 3 tripped at 14:46 JST because of high seismic acceleration. The off-site power source was maintained, but the turbine component cooling seawater pump stopped as a result of inundation by the tsunami. All of the feeding water and condenser pumps were then manually stopped, and the reactor core isolation cooling system fed water to the reactor. Cooling and depressurization operations of the reactor were performed, leading the reactor to a cold shutdown state with a reactor coolant temperature of less than 100°C at 1:17 JST on 12 March.

(a) Flooding due to tsunami runup

At the Onagawa nuclear power plant, the maximum water level observed at the tide gauge (approx. OP + 13 m)[1] was measured about 40 min after the mainshock occurred. The maximum height of the tsunami runup in front of the site where the main building is located was about OP + 13.8 m (see Fig. 38). The ground level of the nuclear power plant site was about OP + 14.8 m before the earthquake but subsided during the earthquake to OP + 13.8 m. The seaside area (OP + 2.5 m) was flooded.

(b) Hydrodynamic forces

Small buildings at lower ground elevation (OP + 2.5 m) near the seaside area were flooded by inundation but largely did not collapse (see Fig. 39).

It was thought that the heavy oil tank was damaged by the tank being floated by buoyancy in the rising water level, causing it to move horizontally and overturn.

---

[1] OP: Onagawa Nuclear Power Plant datum plane for construction (below 0.74 m from T.P.0).

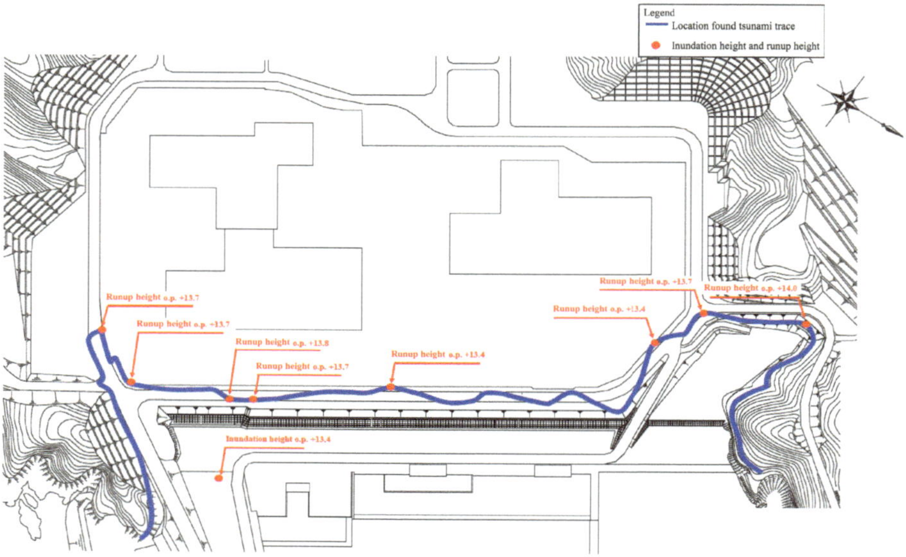

FIG. 38. Tsunami runup in Onagawa nuclear power plant. Reproduced courtesy of Nuclear Emergency Response Headquarters of Japan [62].

FIG. 39. Overturned heavy oil storage tank of Unit 1. Reproduced courtesy of Nuclear Regulation Authority of Japan [60].

(c) Morphological changes

The deposition of sand was between 0.5 m and 0.6 m across a wide area inside the North and East breakwaters; erosion was up to around 5.5 m near the head of the east breakwater. The water intakes were not blocked by the sand (see Fig. 40).

(d) Effects on intake during drawdown

The main tide gauge system was damaged by the tsunami, but the water level of the tsunami was obtained by the backup tide gauge (see Fig. 41). The minimum water level was below OP –6 m for about two minutes, and the intake might have been exposed during this time (see Fig. 42). However, the seawater pump for the emergency component cooling seawater system was able to draw water continuously.

(e) Inundation damage of nuclear facilities

It is presumed that, at Unit 2, the tsunami sea water flowed from the seawater intake channel into the seawater pump room through the installation box of the water level meters. As a result, the reactor component cooling seawater system pump B

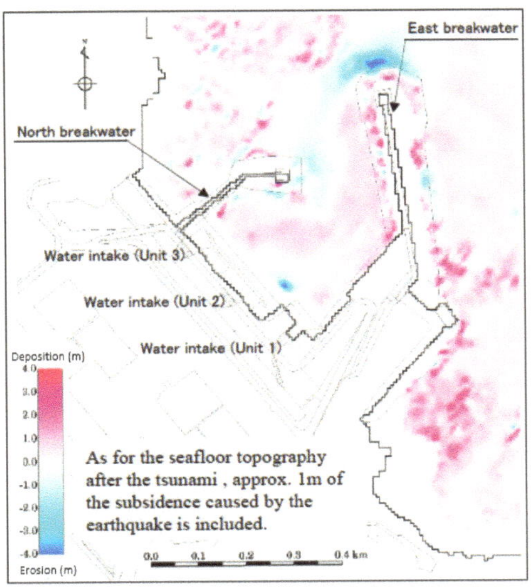

FIG. 40. Morphological change at Onagawa nuclear power plant. Reproduced courtesy of Nuclear Regulation Authority of Japan [60].

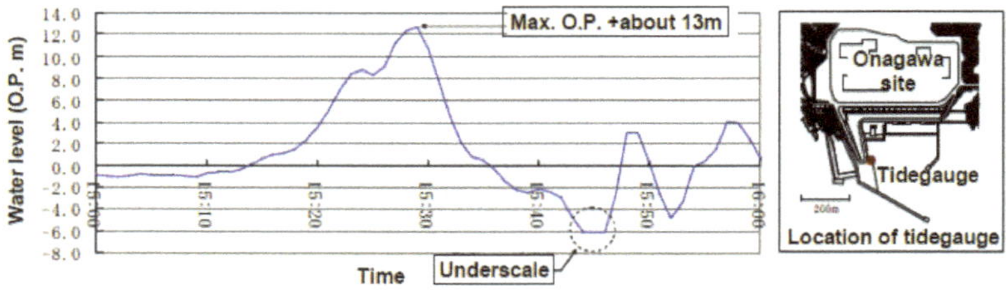

FIG. 41. Time history of the water level change observed at Onagawa nuclear power plant. Reproduced courtesy of Nuclear Emergency Response Headquarters of Japan [59].

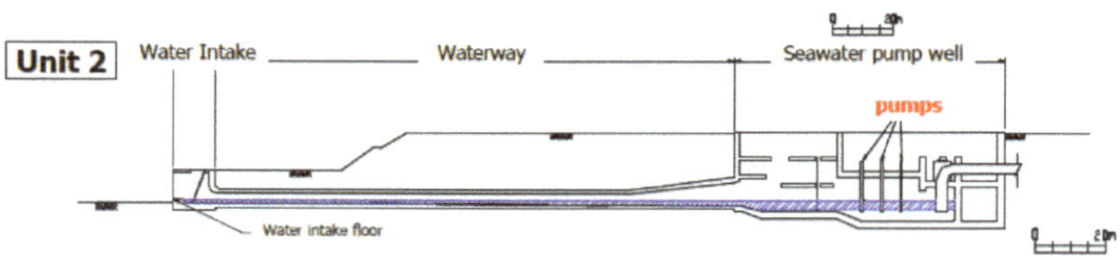

FIG. 42. Water intake system of Onagawa Unit 2. Reproduced courtesy of Tohoku Electric Power Company [65].

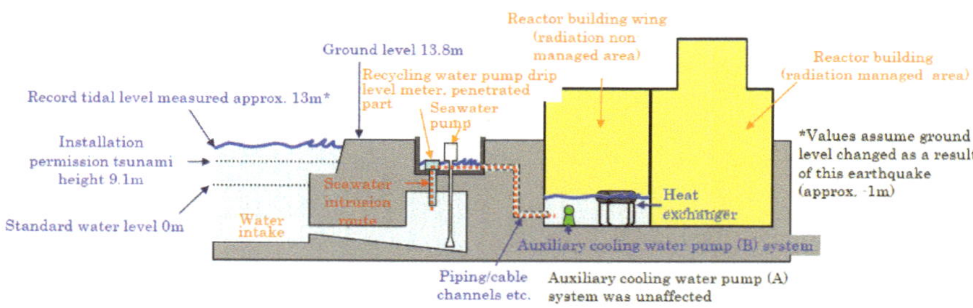

FIG. 43. Penetration pathways into the main buildings of Onagawa Unit 2. Reproduced courtesy of Nuclear Regulation Authority of Japan [61].

region was flooded, and sea water flowed into a part of the reactor building through the underground trench. The reactor component cooling water system B stopped functioning, but system A was intact.

Although water level meters were installed at Unit 1 and Unit 3, they were located in different areas (in the dust arrester rooms), and so sea water did not flow into the reactor building, and the safety equipment was not affected by the tsunami (see Fig. 43).

4.3.3.4. *Tokai Daini nuclear power plant*

There are two stations located near Tokai Village in Naka County, Ibaraki Prefecture (see Fig. 44). The Tokai nuclear power plant started operations in July 1966, with operations ceasing in March 1998, and decommissioning work is being carried out at present. The spent fuel at Tokai nuclear power plant has already been taken out. Tokai Daini nuclear power plant started commercial operation in November 1978. There is a BWR with a generating capacity of 1100 MW(e).

The Tokai Daini nuclear power plant was in full operation at the time of the earthquake on 11 March. At 14:48 JST that day, the earthquake generated a large vibration signal on the turbine shaft. This triggered a turbine trip, which tripped the reactor. Immediately after the occurrence of the earthquake, all three off-site power source systems were lost. However, the emergency power supply to the equipment was secured by the activation of three emergency D/Gs (A, B, C), each of which had three pumps. Subsequently, one of the seawater pumps for D/G (C) tripped as a consequence of the tsunami, and the D/G (C) pump became inoperable. However, the remaining two D/Gs secured the power supply to the emergency equipment, and the cooling of the suppression chamber was maintained by the residual heat removal system B.

One off-site power supply system was restored at 19:37 JST on 13 March, and the nuclear reactor reached a state of cold shutdown with a coolant temperature of less than 100°C at 00:40 JST on 15 March.

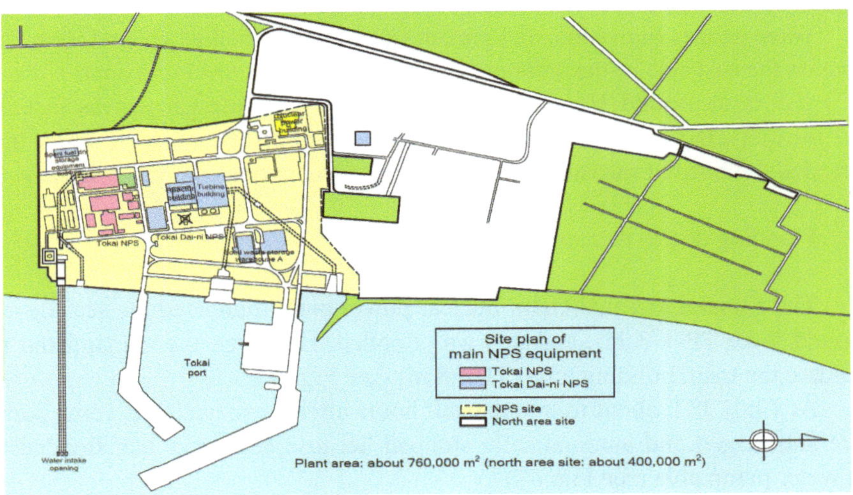

FIG. 44. General layout of Tokai Daini nuclear power plant. Reproduced courtesy of Nuclear Emergency Response Headquarters of Japan [62].

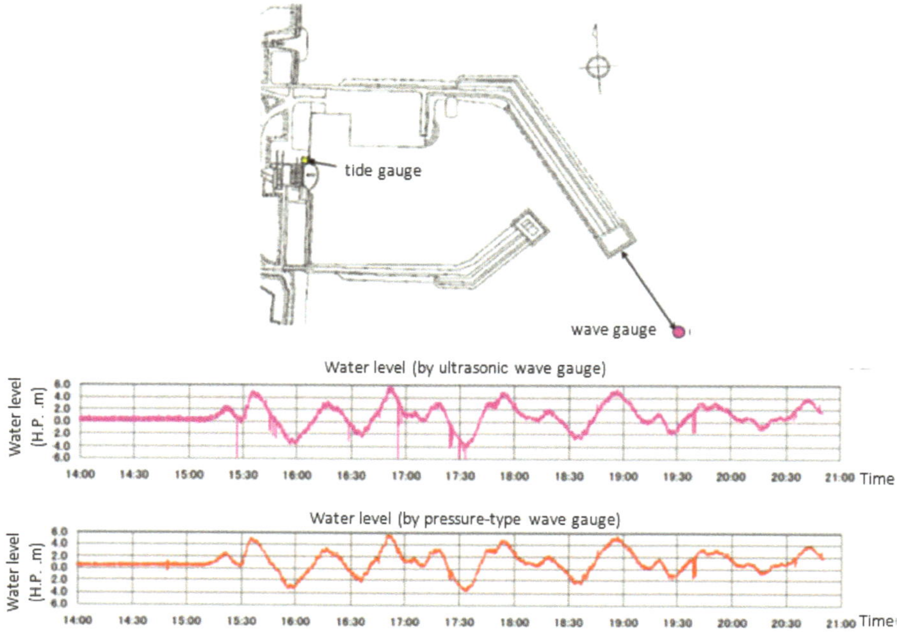

FIG. 45. Water level observed at Tokai Daini nuclear power plant. Reproduced courtesy of the Nuclear Regulation Authority of Japan [60].

Wave gauges (ultrasonic and pressure type) were installed about 150 m from shore on the east side of the harbour. The maximum water level was approximately HP + 5.5 m at around 16:50 JST (see Fig. 45)[2]. The tide gauge was installed near the quay, but the water level could not be recorded because the water level exceeded the limit of measurement, and the electronic power supply was lost.

(a) Flooding due to tsunami runup

The tsunami at Tokai Daini nuclear power plant inundated the seaside area (ground level TP + 3 m), and this was flooded. However, the tsunami did not inundate the main building area (TP + 8 m) (see Fig. 46).

At 19:01 JST, about four and a half hours after the mainshock, some pumps were submerged and automatically stopped because sea water had flooded the sea water pump area (see Fig. 47).

---

[2] HP denotes Hitachi Peil (0.89 m below TP).

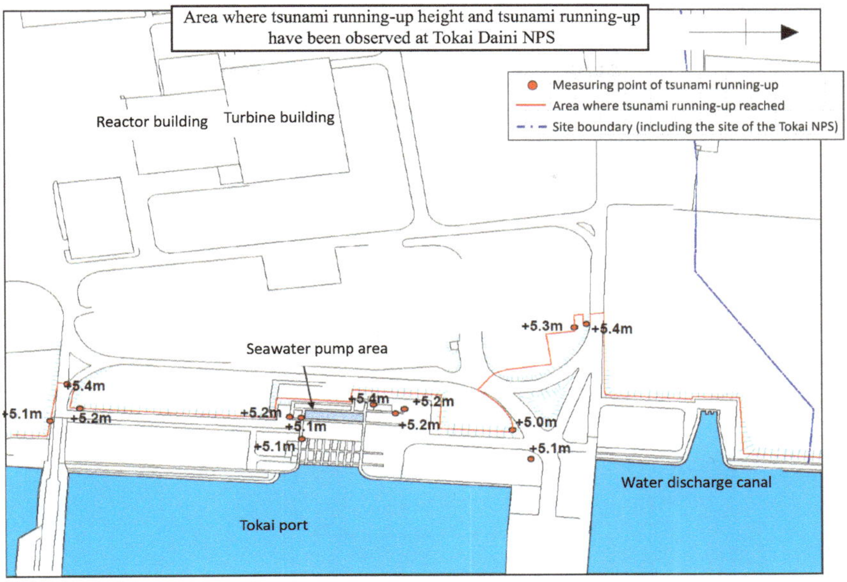

FIG. 46. Tsunami runup (TP) in Tokai Daini nuclear power plant. Reproduced courtesy of Nuclear Emergency Response Headquarters of Japan [62].

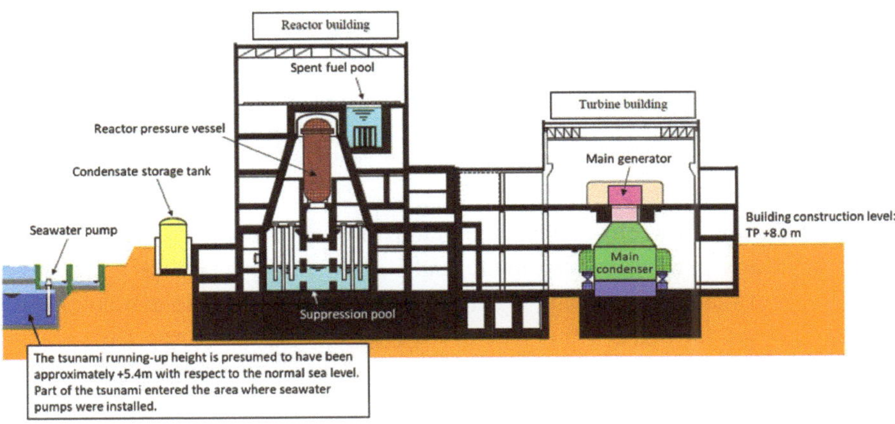

FIG. 47. Schematic cross-section of Tokai Daini nuclear power plant. Reproduced courtesy of Nuclear Emergency Response Headquarters of Japan [62].

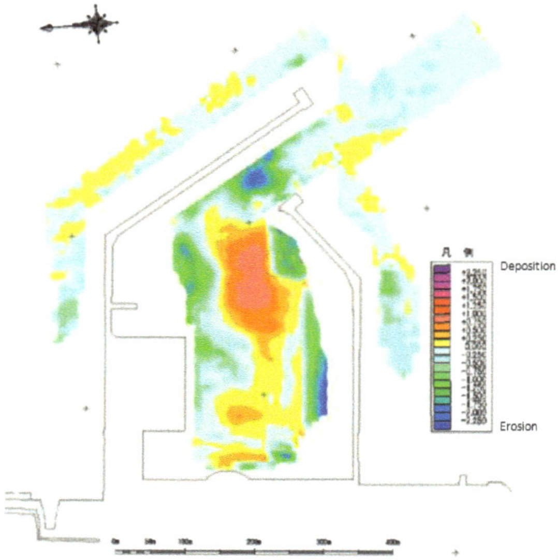

FIG. 48. Morphological change at Tokai Daini nuclear power plant. Reproduced courtesy of Nuclear Regulation Authority of Japan [60].

(b) Morphological changes

Erosion was found around the harbour entrance and inside the harbour near the south breakwater. Deposition was found in the centre of the harbour. No significant deposition was found in front of the water intake (see Fig. 48).

(c) Inundation damage of nuclear facilities

The layout of the water intake pump area is shown in Fig. 49. The circulation water pump is located in the centre of the pump area, with pump areas for equipment that is significant to safety, namely the north side and south side seawater pumps, located at either side of the circulation water pump area.

The new sidewall (TP + 6.1 m), which was installed to improve the safety margin against tsunamis and thus proved to be higher than the tsunami (about TP + 5.3 m), is assumed to have prevented the tsunami from flowing directly into the pump area.

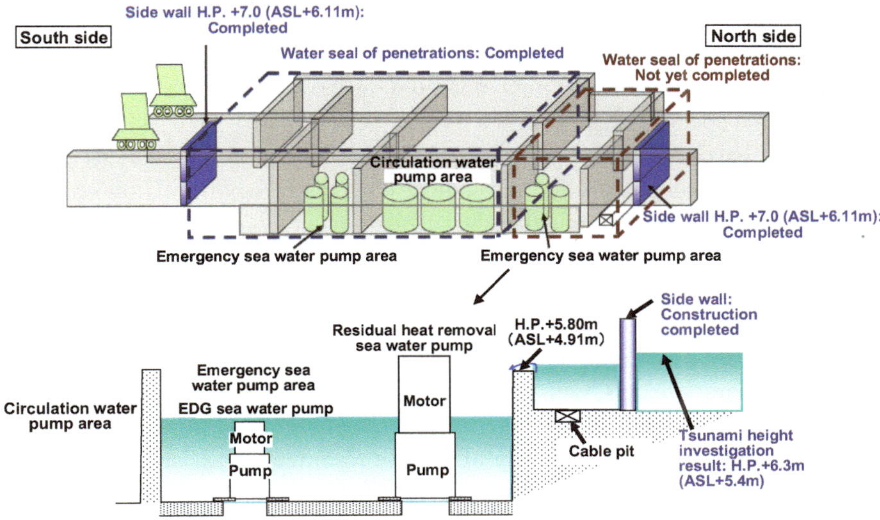

FIG. 49. Inundation to emergency seawater pump area. HP: Hitachi Peil (0.89 m below TP); ASL: above sea level (equal to TP). Reproduced courtesy of Nuclear Emergency Response Headquarters of Japan [59].

Although the tsunami was lower than the sidewall, because of construction work that was going on around the north pump area, sea water could still flow into the pump area at the following places:

— Drain opening between the emergency seawater pump area and the auxiliary seawater system strainer area;
— Cable pit that did not have a waterproof design.

For the south pump area, all activities, including waterproofing of the pipe penetration, had been completed by 9 March 2011.

4.3.4. Lessons learned

Based on the above discussions, the following key points should be considered [66, 67]:

— The importance of a comprehensive analysis of extreme events, such as large tsunamis and seiches and their related uncertainties, in the design and risk assessment of a nuclear power plant within a framework that accounts for margin and defence in depth, especially when limited data are available;

- The uncertainties associated with the potential inundation levels at different locations at a nuclear power plant site due to the plant layout;
- The importance of mapping potential in-leakage pathways for water to inundate specific areas within a nuclear power plant;
- The effects of hydrodynamic forces (the movement of oil tanks, the collapse of a breakwater, etc.), morphological changes (clogging of intake canals) and collision of drifting debris;
- The need to fully and comprehensively investigate the potential of flooding to affect multiple units (and possibly multiple sites) for new and existing nuclear power plants;
- The impact on off-site power sources due to the combined effect of seismic and tsunami impacts, with implications for the extent and timing of expected recovery;
- The impact on the availability of off-site resources to assist in the recovery efforts at an individual nuclear power plant site of (1) extensive damage to the infrastructure around the site, (2) extensive damage to other areas of the coast and (3) the limited availability of emergency resources due to the coast-wide emergency conditions.

5. OVERVIEW AND EVALUATION FLOW

5.1. GENERAL

SSG-18 [1] gives the following guidance regarding the initial assessment of tsunami hazards at nuclear power plants.

> "As an initial assessment, a simplified screening criterion is recommended… Using publicly available information as discussed in para. 3.35, evidence of past occurrences of tsunamis should be reviewed for the site region. For this purpose, the information collected should be organized and a list of specific tsunamis relevant to the plant site should be prepared. No specific further investigations and studies need be performed to analyse the tsunami hazard for the plant site, provided that the site is located in an area that shows no evidence of past occurrences of tsunamis, and is located:
>
> — At more than 10 km from the sea or ocean shoreline, or more than 1 km from a lake or fjord shoreline, as appropriate; or
> — At more than 50 m elevation from the mean water level." (Ref. [1], para. 5.44).

If the above criteria are met, in general, no further investigations and studies need to be performed to analyse the tsunami hazard for the plant site. However, the above thresholds (10 km, 1 km and 50 m) are only empirically recommended limits, generally not reached by known (recorded or experienced) tsunamis. A few tsunamis in the past exceeded 50 m in height, particularly local tsunamis induced by a landslide (NCEI/WDS Global Historical Tsunami Database; see Section A–1). Also, tsunamis can penetrate into rivers by more than 10 km (see Section 4.1.2 and Ref. [68]). Therefore, the above criteria should be considered a general indication and should not be taken as safety bounds, unless the possibility of exceptional tsunami events beyond the given thresholds and long term sea level rise can be completely ruled out on the basis of site and regional characteristics.

On the risk of low water levels, SSG-18 advises the following:

> "In all cases, the required volume of cooling water should be secured in case of the occurrence of a tsunami, because of the potential for low water level to affect the intake water system for several hours" (Ref. [1], para. 5.45).

SSG-18 goes on to describe how a more detailed hazard assessment should be performed:

> "In all situations other than those described in para. 5.44, a detailed hazard assessment for tsunamis should be performed as outlined in the following paragraphs.
>
> "The first step in conducting the detailed assessment of the tsunami hazard at the plant site should be to compile a specific tsunami catalogue and/or database relating to the site. This should be done by the investigations [of geophysical, geological and seismological data] to establish whether or not past or recent tsunami events have occurred in the region of the site and, if so, to characterize them.
>
> "The potential for both local and distant tsunamis should be investigated. The occurrence of underwater and near shore seismic or volcanic activity in the site region (about 1000 km) is an indication of the possible occurrence of local tsunamis at the site. Also, given that large tsunamis can be generated in remote regions, an evaluation of the potential generation of distant tsunamis should be performed for all seismogenic sources existing in and around the specific sea or ocean basin where the plant site is located.
>
> "If the specific studies and investigations performed and compiled in the geological, geophysical, seismological and tsunami databases demonstrate that there is no potential for the occurrence of tsunamis at the site, no further assessment of the tsunami hazard is necessary." (Ref. [1], paras 5.46–5.49 and Fig. 50).

However, the examination of new knowledge on potential tsunami hazards should be continued throughout the lifetime of the nuclear power plant:

> "If, however, a potential for the occurrence of tsunamis at the site is suggested and demonstrated, as a second step, a site specific tsunami hazard analysis should be performed that includes a detailed numerical simulation to derive the design basis tsunami." (Ref. [1], para. 5.50).

When a local tsunami is triggered by a large earthquake, multiple hazards caused by ground shaking and water waves could affect a nuclear power plant site. In this case, a multiple hazards assessment [69] should be conducted.

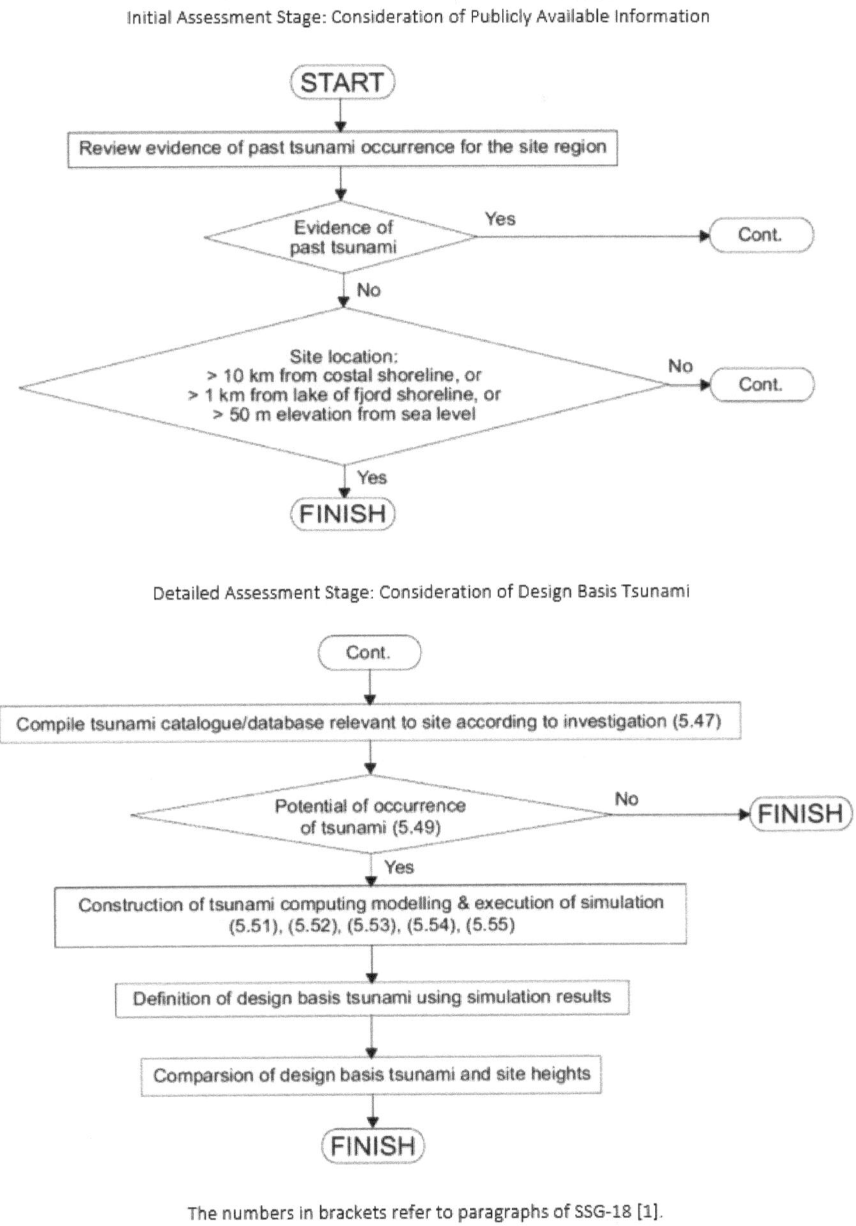

FIG. 50. Flow chart of assessments of tsunami flooding. Reproduced from SSG-18 [1]; the numbers in brackets refer to paragraphs of SSG-18 [1].

5.2. UNCERTAINTIES

On methods for hazard assessment for earthquake induced tsunamis, SSG-18 recommends the following:

> "For earthquake induced tsunamis, the hazard should be assessed by using either a deterministic hazard analysis or a probabilistic hazard analysis, or preferably both methods. The choice of the approach will depend on a number of factors. Whichever method is used, a quantitative estimate of the uncertainties in the results of the hazard assessment should be determined." (Ref. [1], para. 5.56).

These uncertainties are classified into two types: aleatory and epistemic. Aleatory uncertainty relates to the natural or stochastic uncertainty inherent in a physical system. Typically, aleatory uncertainty is represented by a probability distribution from repeated observations or experiments, and it becomes irreducible once a statistically significant number of samples is obtained from the population. Epistemic uncertainty is due to incomplete knowledge (i.e. modelling) and data about the tsunami process. Epistemic uncertainty can be reduced by the collection of new data or the development of more accurate models. Uncertainties in various model parameters, characterizations and assumptions are treated as epistemic uncertainty. The current incomplete knowledge can lead to differences in interpretation by experienced experts on tsunami sources or tsunami runup.

> "Such interpretations should be treated in the tsunami hazard analysis in a consistent manner, providing for a suitable representation of current thinking on tsunami sources, propagation modelling and coastal processes. Particular care should be taken to avoid bias in these interpretations. Expert opinion should not be used as a substitute for acquiring new data. The project team for the assessment of tsunami hazards should not promote any one expert hypothesis or model. It should evaluate all viable hypotheses and models using the data compiled, and then should develop an integrated evaluation that incorporates both knowledge and uncertainties.
>
> "The collection of site specific data tends to reduce uncertainties. However, part of the data that are used indirectly in the assessment of tsunami hazards may not be site specific; for example, the seismogenic data used to characterize the generation mechanism of distant sources. There may therefore be a part of the uncertainty which is irreducible with respect to site specific investigations." (Ref. [1], paras 5.57 and 5.58).

The uncertainties listed below should be taken into account, and both the aleatory and the epistemic parts should be estimated when relevant:

— Uncertainties associated with data (source parameters, bathymetry, topography) (see Section 6);
— Uncertainties associated with the physics adopted to describe the source (e.g. fault models of earthquakes) and propagation (e.g. shallow water equations, non-linear effects) (see Sections 7 and 8);
— Uncertainties associated with numerical calculation (governing equations, grid size, truncation error) (see Section 8).

5.3. CONCEPT OF DETERMINISTIC APPROACH

The design basis tsunami should provide an adequate level of safety to nuclear power facilities concerning all past and possible future tsunamis. In the deterministic approach, this is achieved by estimating the potential tsunami source that would produce a probable maximum tsunami effect at the site, based on historical, geological and geophysical data. It is difficult to treat each uncertainty in a deterministic approach. Furthermore, it is also difficult to select one tsunami source among all the potential tsunamis examined. A systematically conducted sensitivity analysis or parametric study should be used to support the evaluation of the significance of the contributions of the various input parameters in the model.

A parametric study of the dominant factors of the model should be carried out by considering the characteristics of tsunami sources in each region. Section 7 discusses in detail the relevant parameters related to each tsunami generating phenomenon. In earthquake induced tsunamis, for example, the factors for a parametric study should be selected appropriately among the fault position, length, width, depth of upper edge, strike direction, dip angle, slip angle or number of faults. The range of the parameters should be determined from estimated uncertainties. If source model factors can be estimated based on statistical analysis, the range of the parametric values should be adopted considering the standard deviation of the observations with sufficient margin for any limited record of observations.

At a minimum, it should be verified that the maximum runup and maximum drawdown values should envelop those corresponding to historical tsunamis.

Details of the deterministic method are discussed in Section 9.

5.4. CONCEPT OF PROBABILISTIC APPROACH

SSG-18 [1] describes the probabilistic method as follows:

"Probabilistic tsunami hazard assessment is analogous to probabilistic seismic hazard assessment, but it is not the current practice applied by States for assessing tsunami hazards. Methods for the assessment of tsunami hazards using probabilistic approaches have been proposed, although standard evaluation procedures have not yet been developed.

"Results of the probabilistic tsunami hazard assessment are typically displayed as the mean or median annual frequency of exceedance of runup height values through a logic tree approach. The general approach to the assessment of tsunami hazards should be directed towards reducing the uncertainties at various stages of the evaluation process to obtain reliable results driven by data. Experience shows that the most effective way of achieving this is to collect a sufficient amount of reliable and relevant data. There is generally a trade off between the time and effort necessary to compile a detailed, reliable and relevant database and the degree of uncertainty that the analyst should take into consideration at each step of the process." (Ref. [1], paras 5.63 and 5.64).

The outcome of the probabilistic tsunami hazard assessment is an important input to the tsunami probabilistic safety assessment.

Details of the probabilistic method are discussed in Section 9.

6. DATA COLLECTION

6.1. GENERAL

SSG-18 [1] makes the following general recommendations for data collection:

> "When site investigation and data collection are undertaken, care should be taken to include all the information necessary for analysing and estimating site specific values of meteorological and hydrological hazard parameters. All the information collected should be compiled in specific site catalogues or databases for each of the hazards under consideration." (Ref. [1], para. 3.1).

The collected information is related to both the tsunami phenomenon and the sources that generate tsunamis.

> "Detailed studies and investigations should be undertaken to collect all the required and necessary meteorological and hydrological data and information relating to the hazards discussed in this Safety Guide. If it has been conclusively shown in the preliminary investigation that a hazard can be excluded from further consideration, the reasons for doing so should be documented." (Ref. [1], para. 3.2).

> "In all cases, the size of the region to be investigated, the scope and detail of the information to be collected, and the investigations to be undertaken should be sufficient to determine the design bases for protection of the nuclear power plant against meteorological and hydrological hazards." (Ref. [1], para. 3.4).

> "The collection of data and information should be continued throughout the lifetime of a nuclear power plant and up until the completion of the safety related tasks of the decommissioning phase, in order to permit the performance of periodic safety reviews." (Ref. [1], para. 3.5).

> "For the hazard assessment for tsunamis, the available observation periods are generally not sufficient. Other approaches, such as palaeoflood analysis of the site area, should therefore be considered." (Ref. [8], para. 3.8).

The tsunami hazard assessment should be directed towards producing a best estimate result with a defensible range of uncertainties for the consideration of plant safety at various steps of the evaluation process. The degree of uncertainty should be taken into consideration at each step of the assessment. Uncertainty related to tsunami generation includes the pattern of occurrence for tsunami sources. Uncertainty related to tsunami propagation arises from errors or low resolution in the bathymetry and topography. The collection of bathymetry and topography data of higher resolution will lead to a reduction in the uncertainty associated with wave amplification and runup.

6.2. CATALOGUE OF PAST TSUNAMIS

SSG-18 [1] recommends the following regarding data collection for past tsunamis:

> "All data relevant for assessing the potential for tsunami hazards and for determining the parameters of tsunami hazards should be compiled in a tsunami catalogue specific to the site. All historical information and palaeogeological evidence of tsunamis from stratigraphy and other geological studies should be considered in this catalogue." (Ref. [1], para. 3.35).

Available catalogues have been compiled from historical ancient texts, treatises, damage investigation reports, local historical documents and other materials; these are used as basic data when countermeasures against tsunamis and the appropriate location of facilities in coastal areas are under consideration. In historical tsunami studies and tsunami engineering studies, efforts at collecting and accumulating nationwide tsunami trace records have been made. Tsunami trace data are used as important verification data. However, such data include uncertain data (data without a specified reliability level and data with uncertain criteria for measured height). Therefore, in some tsunami trace databases, the reliability of trace data is quantitatively taken into consideration (see Section A–1.2 in the Annex).

The following factors are suggested for consideration:

- — The period for which tsunami reports and records are available.
- — Evaluation of the reliability of the historical documents of each event. This information can be taken from the respective catalogues or publications.
- — Listing of historical tsunami events that have occurred in the nuclear power plant's site region.
- — Estimates of source locations (far field as well as near field), mechanisms, parameters, arrival time, tsunami amplitude, runup and inundation distances,

and damage (if any) for historical events based on the collected data and information for the coastal region of the nuclear power plant.
— Reliability and uncertainty in the reported parameters.

Investigation and collection of seismological data should be undertaken sufficiently for the data to be used in the determination of the tsunami source mechanisms in numerical modelling. These model parameters are used in determining the source of the earthquake induced tsunami as an initial condition to serve as one of the inputs for tsunami numerical modelling.

All available data on submarine landslide and volcanic processes that generate tsunamis should be acquired. All data associated with observed seiches should also be acquired.

The following tsunami databases are available:

— Worldwide Tsunami Database (see Section A–1.1): National Oceanic and Atmospheric Administration (NOAA) NCEI/WDS Global Historical Tsunami Database: http://www.ngdc.noaa.gov/hazard/tsu_db.shtml.
— Historical Tsunami Database for the World Ocean: Institute of Computational Mathematics and Mathematical Geophysics, Novosibirsk, Russia: http://tsun.sscc.ru/On_line_Cat.htm.
— Japan Tsunami Trace Database (see Section A–1.2): Tohoku University and Nuclear Regulation Authority (NRA, Japan): https://irides.tohoku.ac.jp/eng/publication/database/tsunami-db.html.

6.2.1. Prehistorical tsunami deposits

Records of past tsunamis can be found in old documents; however, these span a limited period and usually refer only to inhabited regions. Valuable information regarding prehistorical events or events that occurred in areas not populated in the past, which may greatly expand the catalogue time range, can be found in the geological record buried at coastal sites as typical tsunami deposits. In fact, during tsunami inundation, sediment with different characteristics compared with the underlying layer is deposited widely within the inundation zone. Subsequent alluvial or lacustrine sedimentation may bury these peculiar layers, preserving them locally from alteration and erosion. Over a long period, several tsunami deposits may be recorded at a site. Extensive trench and shallow coring in low lying coastal and near-coastal environments (especially flood plain, lacustrine and marsh areas) can uncover such tsunami deposits and even make it possible to determine their extension inland, which could be crucial to defining the tsunami hazard level at the site. Newly developed chemical, sedimentological and physical methods, added to the traditional dating methods, contribute to better identification of tsunami beds and estimation of the times of occurrence and runup levels of prehistorical tsunami events.

Site and method selection is critical for finding and interpreting tsunami deposits. Coastal lagoons or ponds, where only high energy washover processes can disturb the bottom sedimentation, are the most suitable sites to investigate how often a coast is impacted by tsunamis and when it was last impacted. In such settings, the type of sedimentation commonly allows for high resolution dating, and the column of sediments sometimes spans enough time to record multiple events. Working in these settings is expensive, requiring special techniques for drilling and many stratigraphic and absolute age determinations, and time consuming. Therefore, these studies are not routinely carried out, and so far only sparse local investigations exist, as they are commonly affordable only within the site assessment of relevant plants such as nuclear power plants. Besides, it is possible to obtain only rough estimates (though still valuable) of the size of a palaeotsunami from investigations of lagoon or pond deposits [70].

Identified tsunami deposits are dated using radiocarbon, ^{210}Pb, ^{137}Cs, optically stimulated luminescence and tephrochronological methods. The most commonly used methods are radiocarbon dating and tephrochronology. For radiocarbon dating, the time of tsunami inundation is estimated on the basis of the age of the soil deposited just above and beneath the target tsunami deposit. Where the deposit is found in pure peat, the age determined by accelerator mass spectrometry radiocarbon dating is accurate to a few decades.

To assess the tsunami hazard at sites where tsunami deposits are found, it is critical to know the size of palaeotsunamis at each site. The lateral extent and spatial distribution of tsunami deposits are significantly affected by the local topography of the coastal area, the microtopography of the inundated area and the location and type of the source material. Working in an almost ideal case, where the tsunami has inundated a nearly flat area and no major erosion has followed afterwards (as has often been observed in modern cases), it is possible to identify its areal extent using a tight net of soil profiles. To be of practical use, such a database should be calibrated against the changing coastal morphology, eustatic sea level changes, rates of sediment compaction and tectonic subsidence or uplift.

6.2.2. Historical literature

The tsunami assessment is carried out following overall policies (see Fig. 50). The first step is to conduct old document surveys of the main historical tsunamis affecting the target site and then check the validity of recorded tsunami heights. After the first assessment, tsunami occurrences and the possible effects of historical tsunamis should be listed (see Section 5). If there is no tsunami record for the region, a further tsunami assessment study may still be needed to verify all potential sources. If records of tsunamis are determined in the preliminary assessment step, then a detailed tsunami hazard analysis becomes necessary. Therefore, the following survey should be carried out (see Section 6.2.3).

> "Historical and anecdotal accounts often provide important and otherwise unavailable information that is necessary for improving the comprehensiveness and the reliability of hazard assessments. Care should be taken in both the collection and the analysis of such information. Such accounts are obtained by means of a thorough search of information sources such as, for instance, newspapers, historical records, published and unpublished catalogues of occurrences, personal narratives, runup measurements and inundation zone measurements, field investigation reports, modifications of river channels, film or video records and archives. From data of this type, and by using an empirical classification system for each phenomenon, a set of events and their associated intensities may be collected for the region. Assessments based on these data alone are likely to be biased. This may be due to the scarcity of the data in the range of low intensity events, the dependence of0 the data on the population density at the time (e.g. the phenomenon may have been unobserved in rural regions). The data may also have been subjectively and inconsistently classified at the time, making it difficult to assign an appropriate intensity level to a standard classification method." (Ref. [1], para. 3.9).

Based on the survey results, fault models of historical tsunamis caused by earthquakes can be set up for model verification using numerical simulations. Historical earthquake data may also be used as the basis for probabilistic estimates of tsunamis.

6.2.3. Field survey of tsunami runup and inundation area

When a tsunami event occurs somewhere in the world, expert groups such as International Tsunami Survey Teams visit the affected coast to measure the runup and other related parameters to document the nature and effects of the phenomenon. Much of the data source in such circumstances is perishable and needs to be secured as soon as possible. Similar efforts can also be made concerning the identification of past tsunami deposits. These data can be added to the existing catalogue and used for validation and calibration of source and propagation modelling as well as to make recommendations on the need for further research and numerical simulation.

As an example, Fig. 14 in Section 4.3.2 shows the latest results from all the Japanese Survey Teams for the 2011 Great East Japan tsunami. They include data collected from the Japanese Coastal Engineering Committee groups and other data provided by Japanese prefectural and local authorities. These data, as well as other recent survey data, are included in the NOAA NCEI/WDS Tsunami Event Database (see Section A–1.1) and the Japan Tsunami Trace Database constructed by Tohoku University and Japan Nuclear Energy Safety Organization (JNES)[1] (see Section A–1.2).

6.2.4. Instrumental data

In its recommendations regarding hydrological data, SSG-18 [1] states:

"The water level range for non-tidal phenomena should be obtained, subject to the following considerations:

- Water level records should be obtained for all relevant bodies of water at the site and/or at all gauge stations that are representative of the site conditions for the possible phenomena. The most extensive duration of water level records should be acquired. Attention should be paid to the sampling frequency of data collection to ensure that water level measurements are collected at the appropriate time scale. For example, water level measurements associated with seiche and tsunami related phenomena may be on a time scale of the order of tens of seconds to several minutes [...].
- Wave characteristics (direction, amplitude and period, number of waves and duration) should be reported. Coastal and offshore wave measurements should be obtained using tide gauges, tsunameters or wave buoys, and/or from satellite derived data.
- Field surveys following significant inundation events should include the collection of data on wave height, runup, drawdown and the horizontal inundation, period and duration. Also, the impact of the inundation event on the region (50 km radius) should be collected together with the date, location and information on structures affected (e.g. boats, houses, wharves).
- Water levels for significant historical events [related to phenomena affecting sea surface elevation, such as tides, storm surge, long term sea level rise, etc.] near to the site should be obtained, if available. This includes historical flood marks, tsunami runup heights and historical

[1] Japan Nuclear Energy Safety Organization (JNES) merged into the Nuclear Regulation Authority (NRA, Japan) in 2014. As a result, the Tsunami Trace Database is managed by Tohoku University and NRA.

low water levels during periods of drought. In addition to water levels, other parameters of the inundation (horizontal distance, period), the date of occurrence and the accuracy of the measurements should be reported.

— Special consideration should be given to bore observations. Bores occur in some estuaries, rivers and channels as a result of changes of tide, a tsunami or a sudden change in the discharge through hydraulic structures." (Ref. [1], para. 3.29).

Tide gauge data are available from NOAA, the Intergovernmental Oceanographic Commission Sea Level Web Portal and the University of Hawaii web site:

— http://www.ngdc.noaa.gov/hazard/tide.shtml;
— https://www.ioc-sealevelmonitoring.org/index.php;
— https://uhslc.soest.hawaii.edu/datainfo/.

DART bottom pressure data are available from NOAA and the Nationwide Ocean Wave information network for Ports and Harbours (both GPS and wave gauge data) from Port and Airport Research Institute:

— https://nctr.pmel.noaa.gov/Dart/;
— https://www.mlit.go.jp/kowan/nowphas/index_eng.html.

In addition, note that SSG-18 [1] recommends that "The high tide and low tide levels should be considered in the numerical simulation" (Ref. [1], para. 5.55).

6.3. DATA ON SITE SPECIFIC GEOLOGY AND TSUNAMI SOURCES

On geophysical, geological and seismological data, SSG-18 [1] states:

"Two different sets of geophysical and geological data should be considered with regard to: (a) specific site geology and (b) sources of the tsunami phenomena, if appropriate to the site. The specific geological data that should be collected in the vicinity of the site are data on the following:

— The stability and 'erodibility' of the shoreline;
— Sediment characteristics such as grain size distribution and chemical composition, especially near the water intake structures of a nuclear power plant;

— Hydrogeological characteristics such as permeability and porosity;
— Potential for landslides.

Three types of 'tsunamigenic' sources, both near shore and underwater, should be considered and identified as follows:

— Large 'seismogenic' structures;
— Landslides;
— Volcanic activity." (Ref. [1], para. 3.33.)

Geological and geomorphological mapping should also be considered, and the stability of the seabed should be determined. SSG-18 [1] continues:

"The tsunami source parameters and data on the tsunamigenic potential should be collected for the relevant body of water where the nuclear power plant site is located. The following geophysical, geological and seismological data should be collected for use in determining the source characteristics of potential severe tsunami generators, both local and distant, together with their estimated annual frequency of occurrence:

— For earthquake induced tsunamis: date and origin time, epicentre location, depth, magnitude, seismic moment, focal mechanism (strike, dip and rake angles of the fault plane) and rupture zone parameters (width, length, slip, rigidity, velocity, rising time).
— For landslide induced tsunamis: landslides and cliff characteristics, including location, type and rheology of geological layers, geometry (e.g. slope, size, volume).
— For tsunamis induced by volcanic phenomena: the full characterization of the volcano that may induce tsunamis." (Ref. [1], para. 3.34.)

In the case of landslide induced tsunamis, geotechnical information on slope stability and the rheology of mobilized geological material is needed. Characterization of tsunamis induced by volcanic phenomena may include activity history and types, morphology, likely source mechanisms (flank or caldera collapses, pyroclastic flows, explosions, debris avalanches, etc.), expected dimensions and directions of flow. It must be noted that some mechanisms (e.g. flank collapse and debris flow or avalanche) can be modelled as a special type of landslide source.

6.4. TOPOGRAPHY AND BATHYMETRY DATA

On topographic and bathymetric data, SSG-18 [1] recommends the following:

"The following topographic data should be collected:

- The reference vertical datum and horizontal datum. Special attention should be paid to the possibility that surveys made at different times may have been made using different survey grids or datum. The grid or datum used in each data set should be explicitly stated.
- General topography in the vicinity of the site (to a typical radius of 5 km), with a contour line interval of 5–10 m.
- Detailed topography of the site area and the area immediately surrounding the site that could be flooded, including during the pre-construction and post-construction of the plant.

....

- Elevations and descriptions of levees and other bank protection structures in the vicinity of the site.
- Recent modifications of the topography due, for instance, to a large earthquake." (Ref. [1], para. 3.36.).

Site topography should include satellite images (at 1 m or better resolution). Other factors to consider include bottom friction characteristics on land, including any roughness associated with land use, vegetation, etc., and historical or instrumental evidence of subsidence and uplift. Any breakwaters or sea walls present should be described. SSG-18 [1] continues:

"Bathymetric data to be assembled for the nuclear power plant site should include:

- A common reference vertical datum and horizontal datum for the topographical data.
- Bathymetry of the relevant water bodies, and in particular detailed bathymetry along the shoreline near the plant site. For coastal sites where tsunami or storm surge modelling is proposed, bathymetric data should be assembled... with a spatial measurement interval of no more than 10 m.

....

— Data on long term and short term erosion and/or deposition (from sources such as old surveys, maps, aerial photographs and satellite imagery).
— Recent modification of the bathymetry due, for instance, to a large earthquake." (Ref. [1], para. 3.37).

Bathymetric data should be assembled near the site, depending on the morphological characteristics, and datasets on bottom friction characteristics, including roughness associated with seafloor sediment, textures, etc.

Special attention should be paid to using consistent datums and coordinates when merging topographical and bathymetric datasets. The following aspects may be considered:

— The sets of data should be assessed for outliers and/or morphological changes after data collection.
— All data conversions and transformations should be carefully documented.
— The digital elevation model parameters (extent, cell size, vertical datum, etc.) should be selected such that they satisfy the intended use.
— The existence of artefacts along grid boundaries should be checked while merging mosaics of grids or nested grids. These could arise from a lack of buffer areas surrounding the grid extents, extrapolation, gridding algorithms, etc.
— The accuracy of the coastline data should be cross-checked with that coming from independent data sources.
— The digital elevation values should be verified by visual inspection and in selected cells of the generated digital elevation model with independent data such as geodetic measurements.

The following topographic and bathymetric data are available in the public domain.

— ETOPO: ETOPO is released by the National Geophysical Data Center (NGDC) of NOAA, USA. Over the years, NGDC has released several datasets with increased grid resolution. The first significant release was ETOPO5 (1988), which covered Earth's surface at a resolution of 5 min latitude–longitude grid or about 9 km spacing. In 2006, NGDC released an updated version, ETOPO2v2. Subsequently, ETOPO2v2 was superseded by ETOPO1 (2009). ETOPO1 is a 1 arcmin global relief model of Earth's surface that integrates land topography and ocean bathymetry. See http://www.ngdc.noaa.gov/mgg/global/global.html.

— GEBCO: The General Bathymetric Chart of the Oceans (GEBCO) grid, released in 2008, provides both topography and bathymetry data at a grid spacing of 30 arcsec, which is equivalent to about 900 m. The grid is generated from several sources of data. For seafloor data, this includes data from NGDC, Scripps Institution of Oceanography, USA, the International Bathymetric Chart of the Arctic Ocean (IBCAO) and ship sounding data from French and Japanese agencies. Additionally, some shallow water data (<300 m depth) was also supplied by member states of the International Hydrographic Organization. See http://www.gebco.net/data_and_products/gridded_bathymetry_data.

— SRTM: The Shuttle Radar Topography Mission (SRTM) was a joint project of the National Aeronautics and Space Administration (NASA) and National Geospatial Intelligence Agency. This was accomplished during the mission of Space Shuttle Endeavour in February 2000. The SRTM acquired a digital elevation model of all land between approximately 60° north and 56° south latitude, covering about 80% of the Earth's land surface. The mission extracted the land topography data in a grid of 1 arcsec by 1 arcsec (approximately 30 m by 30 m). The SRTM data are expected to have an absolute horizontal circular accuracy of less than 20 m. Absolute and relative vertical accuracy was anticipated to be less than 16 m and 10 m, respectively. See https://lpdaac.usgs.gov/products/srtmgl1v003/.

— GMRT: The Global Multi-Resolution Topography (GMRT) synthesis is maintained by the Marine Geoscience Data System operating as part of the Interdisciplinary Earth Data Alliance data facility (funded by the US National Science Foundation). See https://www.gmrt.org/about/.

— GTOPO30: Global Topographic Data (GTOPO30) is a global digital elevation model with a horizontal grid spacing of 30 arcsec (approximately 1 km). See http://eros.usgs.gov/#/Find_Data/Products_and_Data_Available/gtopo30_info.

7. SPECIFICATION AND PARAMETERIZATION OF TSUNAMI SOURCES

7.1. BASIC CONCEPTS

7.1.1. Classification of tsunamigenic sources

Tsunamis are caused by submarine geological processes such as earthquakes, landslides or volcanic activities. A different source modelling is used for each type of tsunami. This section describes examples, tsunami source parameters, methods for modelling, validation of the source parameters and other parameters required for deterministic and probabilistic tsunami hazard assessments. The tsunami propagation analysis will be described in Section 8; hence, this section describes the period up to the initial condition of a tsunami from various types of sources.

Table 1 compares the characteristics of the three main types of tsunami sources mentioned above: earthquakes, landslides and volcanic activity.

7.1.2. Concept of uncertainty involved in tsunami sources

Source modelling involves uncertainties. The estimation of source parameters, such as fault parameters or landslide parameters, is usually associated with certain possible ranges. Such ranges, as well as the median values, should be provided for source modelling. The uncertainties can be used for parameter or sensitivity studies in the deterministic method or for branches of scenarios in the probabilistic method. In addition to such random variability or aleatory uncertainties, there are epistemic uncertainties such as the choice of modelling methods. The epistemic uncertainties can be treated as a branch of logic trees in the probabilistic method.

7.1.3. Combination of tsunami sources

Some tsunamis are generated from a combination of sources. For example, subaerial landslides generated by earthquake ground shaking often cause locally large tsunamis. The largest water height in the world, 525 m, was caused by a rockslide with a volume of about 3×10^7 m^3 in Lituya Bay, Alaska, during the 1958 earthquake (M_w 7.9) [71, 72]. During the 1964 Alaska earthquake (M_w 9.2), which caused Pacific-wide tsunamis, local tsunamis from landslides were reported at more than 20 locations, with the largest runup height being 50 m [73].

TABLE 1. COMPARISON OF THE THREE MAIN TYPES OF TSUNAMI SOURCE

	Earthquakes	Landslides	Volcanic activities
Source parameters	Fault model Seismic moment	Subaerial/submarine landslide parameters (volume, velocity, etc.)	Tsunami generation process, duration Displaced water volume
Source modelling	Elastic theory of dislocation	Two-layer modelling	Landslide Caldera collapse Pyroclastic flows Phreatomagmatic explosions
Initial conditions	Seafloor displacement	Water surface displacement Flux	Water surface displacement Flux Water depth
Tsunami model validation	Past earthquakes	Past landslides Experiment	Past volcanic activities Experiments
Deterministic scenarios	Scenario earthquakes	Scenario landslides	Scenario volcanoes
Parameters for probabilistic analysis	Recurrence interval Interevent distribution Size distribution	Recurrence interval Interevent distribution Size distribution	Recurrence interval Interevent distribution Size distribution

This tsunami was recently modelled by a combined source of earthquake fault motion and landslides [74].

Submarine landslides associated with earthquakes or volcanic eruptions also cause large tsunamis. The 1998 Papua New Guinea earthquake (M_w 7.1) caused a devastating tsunami around Sissano Lagoon, with a maximum height of 5 m. However, the large (>5 m) tsunami was limited only to a short (<40 km) segment of the coast, and it was interpreted as being due to additional, presumably submarine slump, sources accompanying the earthquake fault motion [75, 76]. The 1741 tsunami in the Japan Sea originated from Oshima-Oshima, a small volcanic island southwest of Hokkaido, and caused significant damage on the coasts around the Japan Sea, including the east coast of the Korean peninsula.

Recent bathymetry surveys around the island show features indicating a submarine landslide with an estimated volume of 2.4 km^3. The recorded tsunami heights were reproduced by the kinematic landslide model [77] and by the two-layer fluid model [78].

For such combined sources, the parameters described in this section can be jointly used to quantify a tsunami. However, the secondary source of tsunamis, subaerial or submarine landslides, was triggered by earthquake ground shaking or volcanic eruption in the above examples. In such cases, this cannot be treated as an independent source. The probabilities of such secondary or triggered tsunami sources are much higher than the probabilities computed assuming independence.

7.2. TSUNAMIS CAUSED BY EARTHQUAKES

7.2.1. Examples of earthquake induced tsunamis

Earthquakes are the most frequent sources of tsunamis. An earthquake induced tsunami is generated by a seafloor displacement associated with submarine earthquakes of shallow depth (<50 km), large magnitude (M_w >6.5) and dip–slip mechanism. Strike–slip fault motion produces a small vertical displacement of the seafloor, and consequently, the induced tsunamis are usually of small height.

The 2004 Indian Ocean tsunami was caused by the Sumatra–Andaman earthquake with M_w 9.1 at a depth of about 30 km. The fault length is estimated as 1200 km, extending from northwest of Sumatra to the Nicobar and Andaman Islands. The source area of the 2011 Great East Japan earthquake (M_w 9.0, depth 24 km) is much smaller at about 500 km long and 200 km wide (see Section 4). Both events are characterized by a shallow thrust dip–slip mechanism.

Some earthquakes produce a relatively low seismic magnitude relative to the tsunami's size. In other words, the tsunami heights are much larger than expected on the basis of the seismic magnitude. Such an earthquake is called a 'tsunami earthquake' and is distinct from the 'tsunamigenic earthquake', which refers to an earthquake that produces a tsunami. Examples of tsunami earthquakes include the 1896 Sanriku and the 1946 Aleutian earthquakes and, more recently, the 1992 Nicaragua and the 2010 Mentawai (off Sumatra) earthquakes. Tsunami earthquakes usually occur near the trench axis of subduction zones [79, 80].

7.2.2. Parameterization of earthquake induced tsunamis

7.2.2.1. *Static fault parameters*

To express the tsunami source, the fault rupture should be properly modelled to reflect the characteristics of the tsunamigenic earthquake. The spatial distribution of the initial sea level disturbances of the tsunami is generally assumed to be equal to that of the vertical displacement of the sea-bottom by faulting. Further, in the process of estimating the initial sea-bottom displacement, a rectangular fault plane model with a uniform slip has been generally adopted. A large curved fault or subduction interface can be divided into several rectangular faults. A rectangular fault model with a uniform slip can be described by the following nine static parameters (see Fig. 51):

— Horizontal positions of the reference point (latitude and longitude);
— Fault length L;
— Fault width W;
— Slip amount U;
— Depth of the upper edge of the fault plane d_z;
— Strike direction φ;
— Dip angle δ;
— Slip (or rake) angle λ.

Among the above parameters, L, W and U are related to seismic moment (M_0) by the following equation:

$$M_0 = \mu LWU \tag{5}$$

where μ is the rigidity modulus of the medium in the vicinity of the fault. The moment magnitude (M_w) is related to M_0 by the equation:

$$M_w = \left(\log M_0 - 9.1\right)/1.5 \tag{6}$$

where M_0 is given in N·m.

7.2.2.2. *Dynamic fault parameters*

The static parameters in the previous subsection are needed in order to compute permanent seafloor displacement due to faulting. For great earthquakes,

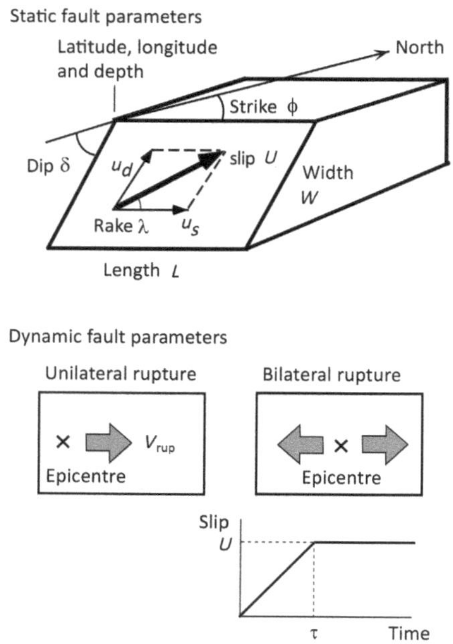

FIG. 51. Fault parameters.

temporal change in faulting may need to be considered. The dynamic or kinematic parameters for describing the temporal variation are the time required for fault slip (rise time) τ, rupture velocity V_{rup} and propagation mode of rupture. The rise time can be set for each point or subfault on the fault plane, or the same values can be assumed. The rupture velocity can be variable on the fault plane, but constant values are often assumed in the kinematic (prescribed) fault rupture models. Rupture modes generally refer to the location of the epicentre or rupture initiation point: bilateral if the rupture starts from the centre or unilateral if the epicentre is located at either end of the fault.

7.2.2.3. Non-uniform fault slip

Seismological and tsunami wave analyses of great earthquakes have indicated that the slip on the fault plane is not uniform but has patches where the large slip is concentrated. Such patches (often called asperities) reflect a strong coupling of the fault for the interplate earthquakes and can be inferred from seismological or geodetic observations. Scaling relations indicate that the ratio of patches to the total fault area is roughly constant for great earthquakes [81]. Many 'tsunami earthquakes' have large slips near the shallowest subfaults close to the trench axis [80, 82].

7.2.2.4. *Rigidity of medium near the fault*

Because the rigidity modulus affects the estimation of M_w from the fault model, it is essential to set an approximate value of the rigidity modulus in the evaluation of tsunamis.

The rigidity modulus of the medium in the vicinity of the fault should be appropriately set depending on the earthquake focal depth or generation area. The rigidity modulus μ can be calculated using the following equation:

$$\mu = \rho V_s^{\,2} \tag{7}$$

where V_s is the S-wave velocity and ρ is the density of the medium. The rigidity can also be estimated from earthquake rupture duration and laboratory testing of rock samples. For the shallow parts of the subduction zone, the shallower earthquakes tend to have a longer duration due to the smaller rigidity [83]. For tsunami earthquakes near the trench axis, smaller rigidity may result in a larger slip for the same seismic moment [80].

7.2.3. Modelling

7.2.3.1. *Coseismic displacement distribution of the sea bottom*

Ocean bottom deformation caused by faulting can be calculated using the elastic theory of dislocation. The displacement u_k in an infinite homogeneous medium due to a dislocation Δu_k across a surface Σ is given by Volterra's theorem (e.g. Refs [84, 85]) as:

$$u_k = \frac{1}{F}\int_{\Sigma} \Delta u_i \,[\lambda \delta_{ij} u_k^{n,n} + \mu(u_k^{i,j} + u_k^{j,i})] v_j d\Sigma \tag{8}$$

where λ and μ are Lamé constants, δ_{ij} is Kronecker's delta, and v is the unit normal to the surface. The expression u_i^j denotes the *i*th component of the displacement due to the *j*th component of point force (F) at the source (Somigliana tensor). For a half space with a free surface, a mirror image can be used to cancel the stress components on the free surface. The explicit formulas are given, for example, in Mansinha and Smylie [86] or Okada [85]. Poisson's ratio v is usually 0.25 (Lamé constants μ and λ are equal), and the formulas by Okada [85] can be used for other values of v.

7.2.3.2. *Duration of fault movement*

The duration of the fault movement that generates a large tsunami is assumed to be approximately from several tens of seconds to several hundred seconds. In most cases, the duration has no significant effect on the results of the numerical simulation of the tsunami as compared with the case in which the sea-bottom is displaced instantaneously (e.g. Refs [87, 88]). Consequently, both methods — with and without consideration of the duration — may produce similar results.

7.2.3.3. *Specifying the initial conditions*

The coseismic displacement of the sea-bottom is usually assumed to be equal to the initial water surface and is used as an initial condition for tsunami propagation. When the dynamic effects of the fault motion are considered in a mass conservation equation, the water surface displacement is assumed to be initially zero. Any initial flux in the horizontal direction (integrated flow in a depth) should be set at zero in either case, that is, with or without the dynamic effects of the fault motion.

The initial conditions given at the ocean bottom and the water surface differ by a factor of $1/\cosh(kd)$, where k is the wavenumber of displacement and d is water depth [87]. If the wavelength of the bottom displacement is much larger than the water depth, this factor becomes unity because $\cosh(kd) = \cosh(2\pi d/\lambda) \approx 1$, hence the water surface displacement can be assumed to be the same as the bottom displacement. If, on the other hand, the wavelength is similar to the water depth, the above factor, which is often called the Kajiura filter, must be multiplied by the bottom displacement to calculate the water surface displacement.

The horizontal displacement of the seafloor may also affect tsunami generation when the seafloor is not uniform but has some slope. The total displacement can be roughly estimated as follows:

$$u = u_x \frac{\partial d}{\partial x} + u_y \frac{\partial d}{\partial y} + u_z \qquad (9)$$

where u_x and u_y are the horizontal displacement in the x and y directions, u_z is the vertical displacement and $\frac{\partial d}{\partial x}$ and $\frac{\partial d}{\partial y}$ are the seafloor slope [89]. For the 2011 Great East Japan tsunami, the horizontal displacement was as large as several tens of metres, hence the effect was significant; 20–40% of the tsunami's amplitude was due to the effect of horizontal displacement and seafloor slope [52].

7.2.4. Verification and validation of the tsunami source model

7.2.4.1. Selection of the target tsunami

Based on the literature survey (see Section 6), the historical tsunami that is assumed to exert the greatest influence on the nuclear power plant site is selected as the target tsunami for evaluation. However, the records of a tsunami that occurred in other regions can be adopted for the validation of the tsunami estimation if no reliable tsunami record can be found around the site.

7.2.4.2. Evaluation of the goodness of fit for the fault model

In general, the fault model that can explain seismic waves is not always the same as the model that can explain the runup of a tsunami. When a fault model for a historical earthquake is set, it is important to set fault parameters that can explain the distribution of the tsunami heights along the coast. However, the reliability of historical data is variable (see Section 6.2.2), and the accuracy of the records must be re-examined on the basis of the original document. If their reliability is determined to be low, they can be eliminated from the comparison.

The goodness of fit of the fault model as the tsunami source can be evaluated by comparing the observed and calculated tsunami heights. Aida [90] evaluated the goodness of fit based on the geometric average K and geometric standard deviation κ. They have been applied as indices of fitness in the space between the recorded and computed tsunami heights. The definitions of K and κ are:

$$\log K = \frac{1}{n}\sum_{i=1}^{n} \log K_i \quad (10)$$

$$\log \kappa = \left[\frac{1}{n}\sum_{i=1}^{n} (\log K_i)^2 - (\log K)^2\right]^{1/2} \quad (11)$$

where n is the number of data points for evaluation, $K_i = R_i / H_i$, R_i is the recorded tsunami height at location i, and H_i is the calculated tsunami height at location i. Since the estimated error of κ depends on the number of samples, the number of samples should be stated for reference in calculating K and κ.

When the fault model of the historical tsunami is considered, the fault parameters are set in such a manner that the recorded runup of the tsunamis along the coast can be satisfactorily explained. In other words, the fault parameters should be set such that the geometric average K becomes nearly 1.0, and the geometric standard deviation κ, whose minimum value is 1, is as small as possible (close to 1).

It is suggested by the Japan Society of Civil Engineers (JSCE) [91] that the following conditions be used as a rule of thumb for K and κ values over wide areas:

$$0.95 < K < 1.05, \quad \kappa < 1.45$$

These conditions are essential to make it possible to explain the overall trend of the runup distribution over a wide area. Further, they are also important for ensuring satisfactory reproducibility in the vicinity of the target site. The following criteria should be considered for selecting runup data to check the reproducibility around the target site:

(a) The distance from the site is short.
(b) The coastal and sea-bottom topographies around the site are similar.
(c) A statistically sufficient number of runups is used for calculating K and κ.

If tide gauge records are used for comparison, the period and/or phase of the tsunami waveforms as well as the maximum amplitude can be used.

When abundant records are available for recent tsunamis, as well as for the earthquakes that caused them, it is possible to refer to various characteristics of the earthquake. These characteristics include aftershock distribution, focal mechanism solution, stress drop, crustal deformation before and after the earthquake, etc.

7.2.5. Specification of scenario earthquake

For deterministic assessment, a scenario earthquake needs to be specified. The scenario earthquake(s) will be chosen from among possible earthquake sources that might affect the site under consideration. Possible earthquake sources include both local and distant sources relevant to the site, and reference will be made to the catalogue of past tsunamis (both historical and prehistorical; see Section 6). Additionally, possible earthquake sources need to be selected from geological and seismogenic structures even when no historical earthquakes are recorded. Such geological and seismogenic structures include plate boundaries (subduction zones) and submarine capable faults.

For the selected source(s), the location of the fault rupture and the maximum possible earthquake magnitude M_{max} need to be estimated. The location can be varied by parametric studies to find the most effective ones for the site. The maximum earthquake size can be estimated by using scaling relationships between the earthquake size M_w and the maximum length L of the possible earthquake source [92].

Regarding the distant source, all the oceanic converging plate boundaries within a reasonable distance need to be considered.

7.2.6. Size and temporal distribution of earthquakes for probabilistic studies

The size and temporal distribution of earthquakes form the principal components of probabilistic tsunami hazard analysis (PTHA). Section 9 describes how other source parameters are determined, many of which can be scaled with earthquake size.

7.2.6.1. Size and frequency distribution

In specific regions, usually larger than the source segment of subduction zones or capable faults, small earthquakes occur more frequently than large ones. The magnitude–frequency relation, known as the Gutenberg–Richter (GR) relation, is written as:

$$\log_{10} N(M) = a - b\,M \tag{12}$$

where $N(M)$ is the number of earthquakes with magnitude M or larger, and a and b are constants. The a value relates to the number of earthquakes with $M \geq 0$ (10^a) and represents the seismic activity in the specified time and region. The b value or slope indicates the relative number of large to small magnitude earthquakes. The b value is typically 1, meaning that the number of earthquakes becomes ten times greater when M decreases by 1, although spatial and temporal variations have been reported. Constant a and b values have been determined for particular seismic zones, such as the Flinn–Engdahl zones (e.g. Ref. [93]) or other seismic zonation schemes (e.g. Refs [94, 95]) using historical and instrumental earthquake catalogues. The Global Earthquake Model project also estimates the magnitude–frequency distributions [96].

While the above GR relation or power-law distribution of earthquakes implies that the earthquake size can be infinitely large, the magnitude has a physical upper limit. The GR relation can be truncated or tapered at a specific maximum or 'corner' M value (see Fig. 52). However, the upper confidence bound of the maximum or corner magnitude cannot be determined from historical data alone. The combination of global earthquake datasets and additional constraints, such as the relative plate motion determined from geodesy, can determine the maximum or corner M_w value and its uncertainty [97].

An alternative hypothesis of magnitude–frequency distribution is the 'characteristic earthquake model' (e.g. Ref. [98]; see Fig. 52). This assumes that there is a maximum magnitude associated with a particular segment, which can be related to the size, the typical length of the segment of subduction zones or capable faults. The frequency distribution may not follow the GR relation over the full range of magnitudes. There might be a gap in the magnitude–frequency distribution, and the maximum (characteristic) magnitude can fall well outside the fitting line. The uncertainty of the characteristic magnitude also needs to be specified.

The distribution of small and large earthquakes in the world's subduction zones, where large tsunamis are generated, may not perfectly fit either GR or characteristic models (e.g. the Cascadia subduction zone [99]).

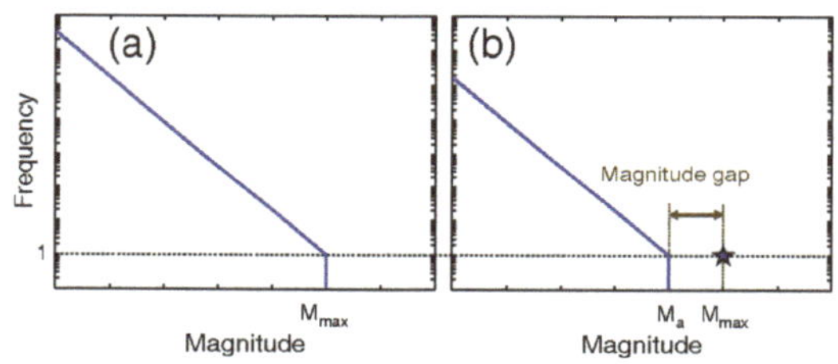

FIG. 52. Magnitude and frequency (number of events) distribution based on Ref. [98]. The left graph shows the GR distribution, and the right graph shows a characteristic earthquake model. Reproduced courtesy of Seismological Society of America [100].

7.2.6.2. Temporal distribution

For the temporal distribution of earthquakes, interevent times are treated as the controlling parameters. Depending on the distribution, the temporal characteristics can be treated as either a quasiperiodic process or a Poisson process. The instantaneous hazard or time dependent probability of instantaneous hazard can be computed from the former, and the stationary hazard or time independent probability can be computed from the latter.

If the distribution of interevent time has a peak, it means that a characteristic recurrence period exists. Such a process is called a quasiperiodic process, and the distribution can be modelled by various probability density functions such as log-normal distribution, Weibull distribution or the Brownian passage time model [101]. It represents the distribution function of the recurrence time of earthquakes with the average μ and the variation α of earthquake recurrence intervals specified as parameters. The variance of Brownian passage time distribution is $(\mu\alpha)^2$.

If the interevent distribution has a power-law distribution without a characteristic peak, it means that the earthquakes occur randomly in time. Such a process is called Poissonian. The probability that a large earthquake will occur during a certain period can be calculated by using the average number of occurrences of large earthquakes per unit of time as the parameter.

7.2.6.3. Consecutive occurrence of earthquakes

Multi-segment earthquakes may occur in interplate events, such as segments along the subduction zones. The probability of this is estimated using historical earthquake data. In the case that tsunami heights change according to the time intervals of multiple ruptures, the tsunami hazard should be estimated by considering the rupture process of the multi-segment earthquake.

The aftershocks of some large earthquakes may be capable of generating tsunamis due to fault break or as a result of triggering a landslide. The 1998 Papua New Guinea tsunami and the 1956 Southern Aegean tsunami are examples of events of this kind.

7.3. TSUNAMIS CAUSED BY LANDSLIDES

7.3.1. Examples of landslide induced tsunamis

There are two broad categories of landslides: (1) submarine or subaqueous landslides, which are initiated and progress beneath the surface of the water body; and (2) subaerial landslides, which are initiated above the water and impact the water body during their progression or fall into the water body. The movement of a large mass of the slide or the impact of the fall displaces the water in the direction of the movement and can lead to the generation of a tsunami wave in the water body. Once the initial wave field has been generated, it propagates outward from the source region.

7.3.1.1. Submarine landslides

Several mechanisms can trigger a submarine landslide. The most common of these is an earthquake, such as the 1929 Grand Banks, 1946 Aleutian, 1964 Alaska and 1998 Papua New Guinea tsunamis. Often, landslides triggered by an earthquake can occur very shortly after the earthquake such that the generated tsunami is affected by both source mechanisms. Many of the events in tsunami catalogues that are attributed to landslides may have such a composite source. In other instances, the slope failure may occur several hours after the triggering earthquake. Submarine landslides can occur in many environments where there is a significant slope of the seafloor, such as continental and island arc slopes, canyons, carbonate platforms and regions with active diapirs.

7.3.1.2. Subaerial landslides

The geographical areas where subaerial landslides occur are more restricted than those where submarine landslides occur. Subaerial landslides, by definition, start on land and then impact a water body. Therefore, their occurrence is generally limited to areas of steep coastal or shoreline topography. One exception to this limitation is debris and other flows that originate away from shore but reach and impact a water body. Examples include the tsunami caused by the 1792 collapse of the southern flank of the Mayuyama dome in front of Mount Unzen, or the tsunami caused by the pyroclastic flow from the 2003 lava dome collapse of the Soufrière Hills Volcano on Montserrat, Lesser Antilles, when it entered the ocean.

The impact velocity of subaerial landslides can be significantly greater than that of submarine landslides. However, typically, subaerial landslides displace less water than submarine slides; thus, the geographical extent of their damage is more limited.

Many of the largest subaerial landslides are triggered by earthquakes, and some by classic hill slope failure mechanisms under wet conditions. For example, in Alaska, USA, destructive local tsunamis have occurred as a result of subaerial landslides, as was the case in Lituya Bay in 1958, and submarine landslides in the Valdez Arm of the Prince William Sound that were triggered by the 1964 Great Alaska earthquake. In southern California, tsunamis have been generated in the geological past by submarine mudflows in Santa Barbara Channel and by debris avalanches in Palos Verdes.

Subaerial landslides have also occurred in inland rivers, lakes — such as the tsunami-like wave in Spirit Lake, Washington, which was caused by debris flow after the eruption and collapse of the Mount St. Helens dome in 1980 — and human-made water storage reservoirs; for example, a tsunami-like wave in Vaiont Reservoir in Italy in 1963 was caused by a massive hill slope failure and the resulting landslide into the reservoir. A survey of tsunamis in Russian inland basins caused by subaerial landslides and other sources is given in Ref. [102].

7.3.1.3. *Ice falls*

These processes include snow and ice avalanches. Ice avalanches are categorized into frontal block failures (calving when the front of the glacier is in water), ice slab detachments and ice bedrock failures (in which part of the bedrock is included in the failure). For example, there exists a historical description of approximately 2 m waves generated by a frontal block failure 7 000 to 16 000 m^3 in volume in a lake in the western Italian Alps.

Ice avalanches moving downslope will behave similarly to initial velocity subaerial landslides, and likely primary source parameters are volume and impact Froude number. Calving will behave like a topple entering the water.

7.3.1.4. *Volcanoes*

Tsunamis can occur as a result of a variety of mechanisms associated with volcanoes that are located in or near oceans or other bodies of water. Volcanic activities such as pyroclastic flows, debris avalanches and flank failures have mechanisms similar to subaerial landslides and avalanches. Tsunami sources related to volcanic activities are discussed in Section 7.4.

7.3.2. Parameterization of landslide induced tsunamis

7.3.2.1. *Source types*

Landslides occur in several ways, depending on the geological composition of the slope, steepness of the slope, triggering mechanism and porewater pressure. There are five classes of slope movement: (1) falls, (2) topples, (3) translational and rotational slides, (4) lateral spreads and (5) flows [103]. Depending on the location, material properties and properties of the trigger, a combination of these movements may occur. These combined movements are called complex slope movements.

The initial tsunami wave generation is affected by the type and the time history of the slope movement. Therefore, it is important to identify these parameters of the landslide in the area of interest. In a given area, several types of landslide events may occur that are capable of generating tsunamis.

7.3.2.2. *Source parameters*

Like the source of the earthquakes, submarine landslides need to be properly parameterized to reflect the essential characteristics of the tsunami generation. The dynamic processes of landslide movement are more important in generating the tsunami than for earthquakes. Also, for translational slides, a downslope moving forcing function (vertical displacement of the water column as a function of time) must be used.

The primary source parameters for a landslide tsunami are the following:

- — Slide location (latitude and longitude);
- — Slide orientation (directivity);
- — Slide dimensions and volume;
- — Density of slide material;
- — Slide speed and acceleration;
- — Cohesiveness and rheology of slide material.

A source parameter for subaerial landslides, including icefalls, in addition to those listed above for submarine landslide is:

- — Impact Froude number.

7.3.2.3. Rotational/translational slides

The most common type of tsunami generating landslide is one that includes some combination of rotational and translational movement, as is schematically shown in Fig. 53.

The region of excavation can involve either rotational or translational movements and result in negative vertical displacement of the ocean surface. The initial acceleration of landslide movement is an important parameter that controls the amplitude of the tsunami trough phase (the leading phase in the backgoing

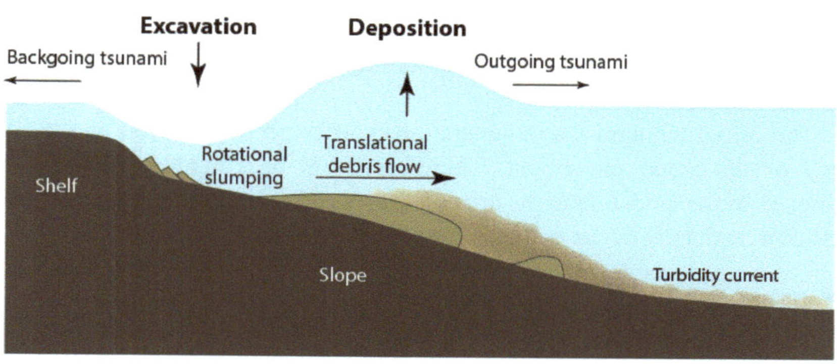

FIG. 53. Schematic of tsunami generation by rotational/translational submarine landslide. Reproduced courtesy of Elsevier [104].

direction). The region of landslide deposition involves primarily downslope translational movement of the debris flow or debris avalanche and results in positive vertical displacement of the ocean surface (see Fig. 53). The speed of this translational movement is an important parameter that controls the amplitude of the tsunami peak phase (the leading phase in the outgoing direction). The tsunami amplitude is 'tuned' in this direction with increased amplification as the landslide speed approaches the tsunami phase speed. The dimensions of the regions of excavation and deposition from past landslides are determined using high resolution seafloor mapping techniques (e.g. multibeam sonar) and seismic reflection profiles.

Portions of the debris flow may separate, particularly during hydroplaning, resulting in outrunner or glide blocks. Turbidity currents are also commonly associated with debris flows (see Fig. 53). However, because they are a bottom boundary layer phenomenon, they have little influence on tsunami generation.

To obtain realistic values for the acceleration and speed of submarine landslides, post-failure, dynamic flow models such as BING (a 1-D numerical code developed by [105, 106]) can be performed. For these models, the parameters of slide density, rheology and yield stress are important. Material parameters for models of landslide dynamics are either obtained from laboratory measurements of samples from the geological material that failed or by trial and error to match the shape and position of the landslide deposit. Landslide dynamic models can be used to determine the parameters of a tsunami generation forcing function (i.e. decoupled) or can be numerically coupled with the tsunami propagation model. Alternatively, static 'hot start' initial conditions based on the difference between the pre-failure and post-landslide bathymetry can be used to conservatively estimate the tsunami generation process.

7.3.3. Modelling

7.3.3.1. Subaerial landslides

Due to the complex dynamics of subaerial landslides and the water waves they generate, most methods involve using either a simplification of the physical processes involved or regression analysis from flume studies or case histories. Thus, the initial tsunami wavefield is usually determined from a slide impact model parameterized using the velocity, density and affective dimension of the landslide. Wave propagation is modelled using the shallow water wave equations, where the wavelength is considered to be significantly greater than the water depth. For subaerial landslides in which the elevation of the failed mass is not great enough to be considered an impact source, methods similar to those described for submarine landslides can be used [107].

A typical example of a tsunami from a subaerial landslide is the 1792 Shimabara tsunami in Japan. In 1792, Mayuyama, a small mountain forming part of the Unzen volcano, collapsed, and the landslide entered Ariake-kai, a shallow bay with a typical depth of 10 to 20 m. Because the horizontal extent of the slide was about 3000 m, much longer than the water depth, this satisfied the long wave approximation, thus causing a large (10 to 20 m) tsunami height and significant damage (with ~5000 casualties) on the other side of the bay. The total number of casualties was ~15 000. For smaller scale subaerial landslides, the shallow water equations may not be adequate, and the non-linear dispersive models should be used as described in Section 7.3.3.2 below.

7.3.3.2. Submarine landslides

Because the tsunami initial conditions for submarine landslides are often smaller in dimension and higher in amplitude compared with the initial conditions for earthquakes, it is important to include the effects of frequency dispersion and non-linearity in propagation models of landslide tsunamis. These effects are described in Section 8. Non-linear, dispersive tsunami models are often sensitive, in terms of numerical stability, to small scale spatial and temporal changes in initial conditions. For this reason, smooth analytic functions are often used to represent vertical displacement changes as a function of time for the landslide tsunami generation process (e.g. Ref. [108]). The functions represent subsidence in the region of excavation and elevation in the region of deposition, as shown in Fig. 53. The functions are parameterized by the dimensions of the slide, a time constant for initial landslide movement and the terminal speed of downslope landslide movement. As for the earthquake source, it may be necessary to apply the Kajiura filter (see Sections 7.2.3.3 and 8.2.2) for small bottom displacement wavelengths relative to the water depth.

7.3.4. Verification and validation of the tsunami source model

Landslide tsunami models that are based on a forcing function can be validated using the results of wave tank studies. For example, laboratory time series and runup values for a landslide of prescribed shape and time history are provided in Refs [109] and [110]. Numerical tsunami models have also been validated by large scale wave tank experiments using freely sliding wedge shaped blocks [111] (see also the NOAA Center for Tsunami Research web site[1]). The goodness of fit for numerical models of landslide tsunamis can be evaluated using the statistical measures described in Section 7.2.4.2.

---

[1] https://nctr.pmel.noaa.gov/benchmark/Laboratory/Laboratory_Landslide/index.html

If numerical models of landslide dynamics such as BING are used to constrain the time evolution of the tsunami generation forcing function, the final shape and position of the tsunami deposit are often used to evaluate the performance of the model. Examples include studies of the Palos Verdes debris avalanche [112] and the Currituck landslide [113].

7.3.5. Specification of scenario landslide

As for earthquakes, a scenario landslide needs to be specified for deterministic hazard assessment. The scenario landslide is chosen from a survey of submarine landslides near the site, in which the type (rotational, translational, etc.) and maximum dimensions and size are determined. The offshore survey is conducted using high resolution seafloor mapping and seismic reflection profiling techniques. If no significant submarine landslides are found near the site, landslides at regional and far field distances should be considered.

After surveys of landslides near the site have been conducted, it should be determined whether present-day geological conditions offshore of the site are similar to those for the surveyed landslides. Surveyed landslides may have occurred tens of thousands of years ago under very different sea level conditions and with sediment input from glacial activity, for example. The maximum size landslide of those surveyed is used as the scenario landslide if that landslide did not occur under drastically different geological conditions.

7.3.6. Size and temporal distribution for probabilistic studies of landslides

As with earthquakes in Section 7.2.6, the size and temporal distribution of landslides are described here for PTHA.

7.3.6.1. Size distribution

The distribution of landslide sizes (i.e. areas or volumes) near the site can be determined using offshore geophysical surveys, such as multibeam mapping of the seafloor. To calculate the sizes of many landslides, GIS (Geographical Information System) methods can be employed, in which the perimeter of the excavation region is digitized, and a surface is automatically fitted to determine the area (see Fig. 54). The difference between this surface and the seafloor in the excavation region approximately determines the volume [114]. The landslide sizes are then binned to determine the empirical distribution function.

Only recently has there been sufficient information to determine the expected size distribution of submarine landslides. In some cases, the distribution of landslide volumes follows a power-law relationship like their on-land

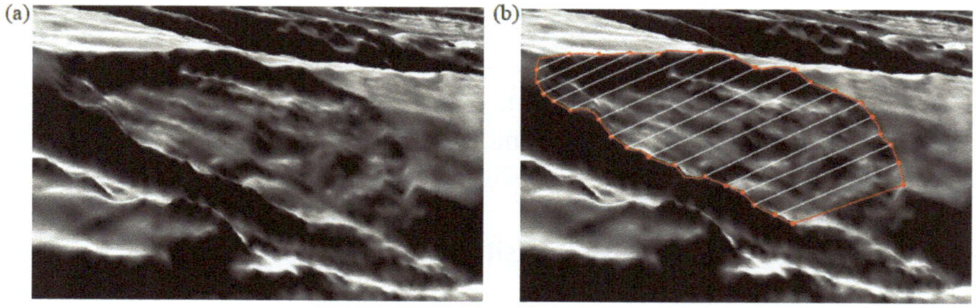

FIG. 54. Determination of the distribution of landslide sizes [114]. Left: a perspective view of the landslide excavation region from multibeam bathymetry. Right: the red line and points show the perimeter of the excavation region to determine the area (white shaded region) and volume of the landslide. Reproduced courtesy of Chaytor et al. and Elsevier [114].

counterparts, modified to include a taper or truncation at the largest sizes. An example is from the carbonate platform offshore of Puerto Rico [115]. In other cases, such as the clastic continental shelf offshore of the US Atlantic margin, a log-normal distribution of landslide volumes is more appropriate [114]. For any given region, different distribution models should be statistically tested to determine which is the most appropriate model, given the data.

7.3.6.2. Temporal distribution

There is typically very little data to determine the temporal distribution of submarine landslides. As a null hypothesis, one can assume that landslides occur independently throughout time as associated with a Poisson process and an exponential distribution of interevent times.

Ideally, in order to determine the temporal distribution of submarine landslides, a sequence of submarine landslide deposits imaged using seismic reflection data would be drilled or cored. Geological age dating of strata above and below each landslide deposit would constrain the absolute age of each event (e.g. Ref. [116]), from which an empirical distribution of interevent times could be determined. Other stratigraphic techniques using nearby age controls could also be used to determine the age of each landslide event.

If the age dates of each event are not available, but the age of a basal seismic horizon below a sequence of landslides is known, then it could be possible to determine the rate parameter and uncertainty, assuming an exponential distribution associated with a Poisson process [117]. If no age data are available, indirect techniques may be used that link seismic activity to a geotechnical slope failure model (e.g. Ref. [118]). In this case, it would be necessary to keep track of all uncertainties associated with the model and incorporate them into the probabilistic analysis.

Because the occurrence of submarine landslides appears to be dependent on climate, glaciation and sea level, the rate of occurrence may be non-stationary; that is, the rate may vary throughout geological time. If there is sufficient age data, changes in the rate over the time in which the surveyed landslides near the site occurred need to be accounted for to accurately determine the present-day rate of occurrence.

7.4. TSUNAMIS CAUSED BY VOLCANIC ACTIVITY

Compared to the total number of tsunamis listed in the available catalogues (more than 2000 events in the NCEI/WDS Global Historical Tsunami Database[2]; see Section A–1.1), the number of events somehow related to volcanic activity is relatively few, the NOAA database containing 131 of them out of a total of 2145 (see Fig. 55). Moreover, the volcanic source mechanism is particularly difficult to characterize, and several of these events would probably be better listed under the landslide or earthquake source (see Sections 7.2 and 7.3). However, the volcanic tsunamigenic source is far from a minor one, having produced some of

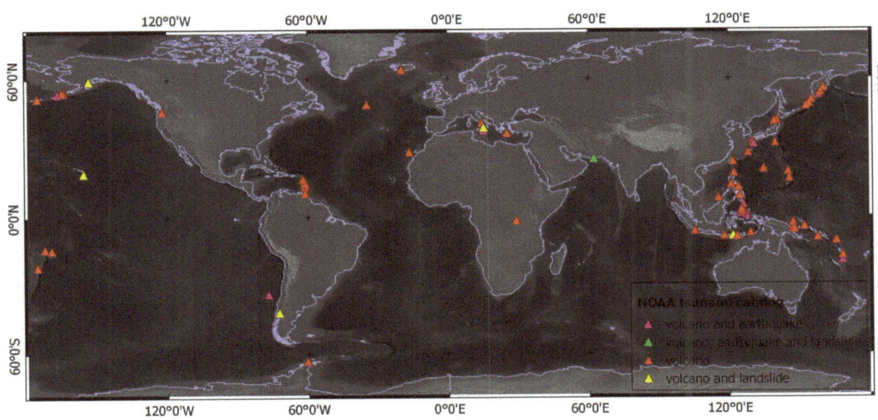

FIG. 55. Volcano related tsunamis in the NCEI/WDS Global Historical Tsunami Database. The events are mostly focused on the Pacific 'Ring of Fire', Hawaii, the Caribbean and the Mediterranean Sea.

---

[2] http://www.ngdc.noaa.gov/hazard/tsu_db.shtml

the largest known events on Earth, such as the Santorini event in 1650 BCE and Krakatoa in 1883 CE. Gigantic slope failures from the flanks of island volcanoes may cause devastating tsunamis even in areas less prone to this hazard, such as the Atlantic coasts (e.g. Ref. [119]).

7.4.1. Examples of volcano induced tsunamis

Volcano edifices are typically unstable structures, and so any volcano located near or under water is a potential source of tsunamis. Magmatic explosive eruptions, phreatomagmatic explosions, gas bursts, caldera collapses, flank failures or magmatically induced seismic activity of submarine, island or coastal volcanoes are typical causes of tsunamis (e.g. Refs [120, 121]). The most frequent mechanisms are associated with slope instabilities that are directly connected to paroxysmal volcanic activity, namely the rapid collapse or slide of a portion of the volcano edifice (originating either in a submarine or subaerial environment). Bathymetric surveys reveal that shield volcanoes in oceanic settings have been the site of huge submarine debris avalanches, which could have potentially resulted in basin-wide tsunamis (e.g. Ref. [122]). Landslides occurring on volcanic islands are among the largest mass movements on Earth, able to reach volumes of up to several hundred cubic kilometres, according to some authors (e.g. Refs [123, 124]). When sufficiently fast, such slides can generate devastating tsunamis [125–127] and represent a significant hazard from the local to regional scale (e.g. Refs [128–130]). However, hazardous amplitudes are rarely maintained on an ocean basin scale (e.g. Ref. [131]), as is often the case for the major subduction earthquake tsunamis. Even moderate eruptions of island volcanoes may generate notable, although much more localized, tsunamis (e.g. Stromboli volcano in 2002 [132]). Furthermore, volcano slopes can become unstable and collapse without warning or eruptive activity.

Pyroclastic flows during major explosive eruptions, superficial or submarine phreatomagmatic explosions and gas bursts can also displace huge volumes of water, causing locally destructive tsunamis. Moreover, although quite rare, one of the most hazardous mechanisms generating volcanic tsunamis is associated with caldera collapse, which consists of a sudden collapse of a volcano edifice, which may be many hundred metres high and kilometres wide, to produce a marine caldera. In such a case, enormous masses of water are displaced because of the rapidly falling volcano edifice and are attracted toward the newly born cavity. The eruptive episodes of Santorini in the Aegean Sea (1650 BCE) and Krakatau in Indonesia (1883 CE) were followed by collapses, which could have generated basin-wide tsunamis that impacted coasts and harbours quite far from the source volcano.

Table 2 summarizes the main tsunamigenic processes that are directly associated with volcanism, source volumes and tsunami characteristics obtained from a review of historical tsunamis generated by volcanic eruptions [133, 134].

TABLE 2. TSUNAMIGENIC PROCESSES ASSOCIATED WITH VOLCANISM AND CHARACTERISTICS OF PRODUCED TSUNAMIS [134]

Mechanism	Source volume (km^3)	Wave height (m)	Period (min)	Travel distance (km)
Earthquakes	1–10	Up to 17	10–40	<500
Submarine explosions	<1	1–6	1–10	<50
Pyroclastic flows	1–100	Up to 25	1–40	<250
Caldera collapse	1–10	Up to 15	Short	<50
Avalanches of cold rock	<1	1–10	Short	<50
Basal surges and shock waves	<1	Up to 5	Aperiodic	<10
Avalanches of hot rock	<1	Small	Short	<10
Lahars	<1	Small	Short	<50
Atmospheric phase coupling	?	Small	15–40	>1000
Lava entering the sea	<1	Very small	Short	<10

For a contemporary review of volcanic tsunamis, Ref. [135] includes a catalogue of events in Southeast Asia and a wide list of references. Figure 56 shows schematically the most relevant tsunamigenic sources related to volcanism [135].

The most relevant historically recorded tsunamigenic volcanic eruption is perhaps that of the Krakatau volcano in the Sunda Strait in 1883 [136]. This event resulted in a tsunami with a maximum runup of approximately 42 m on Java and Sumatra and reportedly caused tsunami damage as far away as Western Australia and India. A total of 34 000 fatalities occurred in Indonesia as a result of the eruption, nearly all related to tsunamis. One lesson to draw from this event is that volcanic activity can result in complex sequences of events and multiple episodes of tsunami formation [120, 133, 137]. The tsunamis during the 1883 Krakatau activity were generated by a variety of mechanisms, including the collapse of the northern part of the volcanic island in the Sunda Strait, caldera subsidence, pyroclastic flows entering the sea and underwater explosions. These events occurred over several days, and tsunamis appear to have been related to each of them.

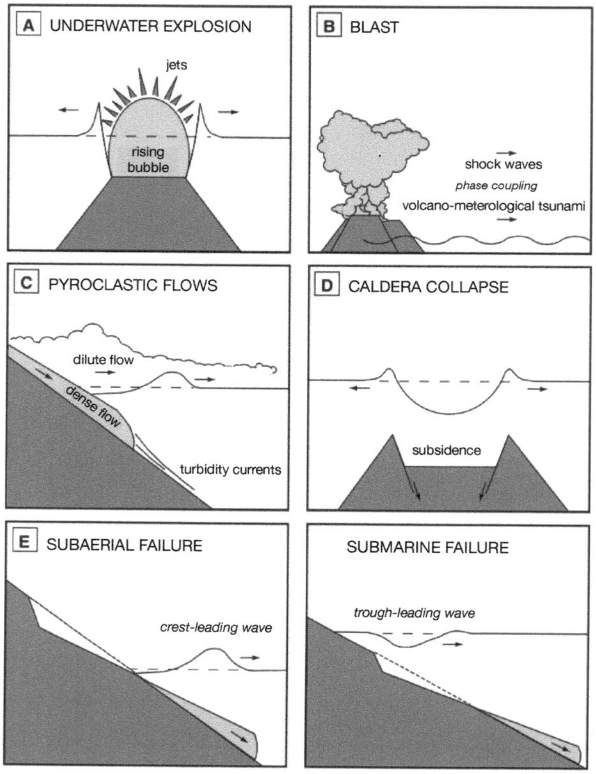

FIG. 56. Schematic mechanisms of tsunami generation related to volcanic activity. Reproduced courtesy of Springer Science+Business Media B.V. [135].

The large tsunami of 1741 in the Japan Sea caused significant damage to the Japanese as well as Korean coasts. The tsunami originated from the Oshima-Oshima volcano, whose eruption was followed by a large scale sector collapse. While the subaerial landslide volume was only 0.4 km^3, the total volume including the submarine part was 2.4 km^3, large enough to produce the recorded tsunami [77].

Examples of some historical tsunamis that have been generated by the mechanisms listed above are provided in Table 3.

TABLE 3. EXAMPLES OF HISTORICAL VOLCANO TRIGGERED TSUNAMIS AND THEIR MAIN CHARACTERISTICS

Event	Year	Mechanism	Volume (km^3)	Max. runup (m)	References
Aniakchak, Alaska	3.5 ka	Pyroclastic flows	~15	~30	Waythomas and Watts [138]
Kikai, Japan	7.3 ka	Caldera collapse, pyroclastic flows	~25	<20	Maeno et al. [139]
Krakatau	1883	Submarine explosion, pyroclastic flow, caldera collapse	5–12	30–40	Maeno and Imamura [140]
Matavanu, Samoa	1906	Lava flow		3.5	Power and Downes [121]
Mount St. Helens	1980	Landslides/debris avalanches into Lake Spirit	2.5	200–400	Voight et al. [141, 142]
New Hebrides	1878	Volcanic earthquake		17	Power and Downes [121]
Ritter Island, Papua New Guinea	1888	Sector collapse	4–5	15	Ward and Day [143]
Soufrière Hills, Montserrat	1997	Landslides/debris avalanches	2–2.5	15	Heinrich et al. [144]
Soufrière Hills, Montserrat	2003	Dome collapse, pyroclastic flows	0.2	15	Watt et al. [131]
Stromboli, Italy	2002	Volcanic landslide		11	Tinti et al. [132]
Unzen, Japan	1792	Sector collapse	0.3	20	Aida [145]

7.4.2. Parameterization of volcanic source induced tsunamis

For most of the volcanic sources, there are inherent difficulties in the reliable assessment of their descriptive parameters, which limit the actual significance of any modelling effort. As already mentioned, volcanic eruptions are generally complex events, and even a single eruption can result in multiple episodes of tsunami formation, possibly influencing each other.

7.4.2.1. Tsunamis generated by landslides

Apart from their peculiar triggering mechanisms and lithological or sedimentological characteristics, most of the volcanic landslides can be treated similarly to any other landslide source [145, 146]. Therefore, see Sections 7.3.2 and 7.3.3 for a detailed illustration of parameterization, modelling techniques and issues related to this type of source. Parameters to be considered are those describing the causative volcanic event and those characterizing size, lithological composition, the cohesion of the sliding mass, rheology, kinematics, the geometry of the sliding surface, Coulomb bottom friction and resistance force due to drag and shear stresses exerted by the ambient fluid, and the time history of the driving parameters from the onset of displacement to its arrest (see also Section 7.3.2).

For volcanic debris avalanches, available data [141, 142, 146, 147, 148] suggest that high impact velocities are easily attained, thereby generating supercritical tsunamis: for water depths less than 100 m, landslide velocities greater than 31 m/s would already be sufficient to yield $F_r > 1$ (see illustration of Froude number in Section 7.3.2). Landslide dynamics cannot be well constrained for large volcanic flank failures, but the most catastrophic scenario appears to be a landslide moving as a single, coherent body [149] or the case of Mount St. Helens, where 2.7 km^3 of material slid from the top sector of the volcano, evolving into an enormous debris avalanche entering Spirit Lake at a speed possibly exceeding 200 km/h, thus causing a gigantic wave which rose to more than 250 m (e.g. Refs [141, 142]).

7.4.2.2. Tsunamis generated by caldera collapse

Hazardous tsunamis can likely be originated by caldera collapse during large explosive volcanic eruptions in a near marine environment. The resulting sea level changes may create more potential energy for tsunami generation than other possible mechanisms such as pyroclastic flows entering the sea or phreatomagmatic explosions (e.g. Ref. [139]). The approach used by Maeno et al. [139] requires knowledge of the size and geometry of the displaced volume resulting from the topographic changes (i.e. water depths before and after caldera collapse) and the duration of the collapsing process. Critical information, such as the mode and speed of caldera collapse, is also a key factor but is difficult to constrain in tsunami generation (e.g. Ref. [139]).

7.4.2.3. *Tsunamis generated by pyroclastic flows*

Another mechanism able to generate tsunamis is the large discharge of pyroclastic material into the sea. Required parameters are the emitted volume of material, the discharge rate and the density of the pyroclastic mixtures (which also controls the mode of travel, i.e. floating on the sea surface or submerged, depending on whether the material is lighter or denser than sea water).

7.4.2.4. *Tsunamis generated by phreatomagmatic explosions*

Tsunamis can be generated by underwater explosions (e.g. Refs [150, 151]). A near water surface explosion may produce a water crater and even expose the seafloor to the atmosphere. After reaching its maximum height, the water crater collapses, and, under the influence of gravity, the water quickly flows inward to fill the gap. Therefore, important parameters are water depth, the dimensions and the shape of the mobilized mass of water, which is proportional to the explosion energy (measured in joules). The explosion energy might be estimated based on the empirical relationships between crater size and explosion energy (e.g. Ref. [152]).

7.4.3. Modelling

Although a few methods have been proposed for the modelling of tsunamis induced by volcanic phenomena, standard evaluation procedures have not yet been developed. In the following, a short overview of recent research results in this field is provided, underlining that, despite the continuous progress in modelling techniques, there are inherent difficulties in the assessment of controlling parameters for most of the volcanic sources that limit the actual value of any modelling effort. The latter is less valid for the volcanic landslide source, whose methods and limitations are the same as those already addressed in Section 7.3.3 for the generic landslide source.

In terms of complex volcanic sequences of tsunami generating events, the first stage of tsunami modelling, treating each source as an independent event, is sufficient to describe the first order behaviour of the processes. However, this involves describing the complex interactions of all the processes mentioned above and is often difficult to predict and reproduce, even using state of the art modelling. Therefore, to be reliable, modelling should account for evolving and interacting phenomena without using drastic simplifications [139, 140], which requires further scientific investigation and effort.

7.4.3.1. Tsunamis generated by landslides

The approaches commonly applied are illustrated in Section 7.3.3. As reported there, the main difficulty is the modelling of landslide propagation. Volcanic landslides often exhibit particularly complex behaviour that is currently impossible to describe physically in a robust way (see Ref. [146]). To demonstrate the application of their model, Kelfoun et al. [149] have simulated landslides entering the sea from the eastern flank of Piton de la Fournaise volcano (La Réunion Island), considering various possible volumes (from 0.5 to 10 km^3), and analysed the impact of the tsunamis generated in this way using a modified version of the VolcFlow program developed by K. Kelfoun at the Observatoire de Physique du Globe de Clermont-Ferrand.[3] The program, which is specifically devoted to the simulation of dense isothermal volcanic flows, also makes it possible to model tsunamis generated by the entry of a rock avalanche into a water body, like those typical along volcano flanks, a modern-day example of which is the 1980 event at Mount St. Helens cited before. Another example of such modelling is that presented by Giachetti et al. [127] about the Güímar debris avalanche (Tenerife, Canary Islands). Modelling by Ward and Day [119] of the collapse of a huge volume (150 to 500 km^3) of the west flank of the Cumbre Vieja Volcano in La Palma Island (Canary Islands) shows potential for wave heights from 3 to 25 m on the coasts of the Americas. The failure may start from the crest of the volcano at a detachment fault that was activated in 1949 (rupture 4 km long, offset 4 m) and has remained inactive since then. In Ward and Day's model, wave height depends largely on the speed of mobilized material.

7.4.3.2. Tsunamis generated by caldera collapse

Tsunamis generated by caldera collapse are commonly simulated using a simple piston-like plunger model combined with a single layer shallow water model (e.g. Refs [139, 140]). The process is mainly controlled by topography changes due to caldera collapse, that is to say the topography existing before and after the collapse, as well as collapsing speed. Maeno et al. applied this model to examine tsunamis during a caldera formation at Kikai caldera in Japan [139] and to explore the most plausible mechanism of the tsunami generated during the 1883 Krakatau eruption in Indonesia [140]. Geometrical changes were introduced

---

[3] https://lmv.uca.fr/volcflow/

using a time dependent water depth in the collapsing area, *d*(*t*), which is assumed to fall either with a uniform velocity or following free fall conditions:

$$d(t) = d_{before} - t(d_{before} - d_{after}) / t_s \tag{13}$$

$$d(t) = d_{before} - 1/2gt^2, for\ t < t_s \tag{14}$$

where d_{before} and d_{after} denote the still water depths before and after caldera collapse, respectively, t denotes the time, t_s the duration of collapse and g the gravitational acceleration.

Typically, the duration of caldera collapses is not well constrained. For the case of the Kikai caldera, rapid collapse conditions with durations of a few minutes to a few tens of minutes were able to generate the largest tsunamis [139], although the most plausible collapse duration was estimated to be longer than several hours [139, 153]. Caldera collapse speed V_c can be used to calculate the dimensionless collapse speed $V_c / \sqrt{gd}$. The computed maximum height of the tsunami is greatest when the dimensionless collapse speed $V_c / \sqrt{gd}$ is about 0.01, and the height substantially decreases with slower speeds (longer collapse durations). Simulations for Kikai caldera showed that when the collapse speed was high, the sea level rose rapidly, and a large tsunami was generated because a large amount of sea water flowed into the collapsed area in a short time and the wave crest easily became higher than the original sea surface. Based on the results of the numerical simulations, Maeno et al. [139] suggested that when $V_c / \sqrt{gd} > 0.003$, a tsunami could inundate the investigated area.

7.4.3.3. Tsunamis generated by pyroclastic flows

Maeno and Imamura [140] simulated tsunamis generated by pyroclastic flow using an approach similar to that adopted by Kelfoun et al. [149] for landslide generated tsunamis. In particular, they consider a two-layer shallow water model for two types of fluid (i.e. a dense type and a light type). Shallow water equations of mass and momentum continuity in each layer are solved by considering kinetic and dynamic conditions at the free surface and interfaces. Moreover, the model assumes a hydrostatic pressure distribution and negligible interfacial mixing, with uniform velocity, density and distributions, although a density change by particle sedimentation can eventually become significant with time. The two-layer model is used in the near field, whereas a single layer model is used in the far field. A dense type two-layer shallow water model is assumed for pyroclastic flows denser than sea water. In this case, dense pyroclasts are assumed to be the dominant components of the flow, which can therefore enter into sea water and travel along the slope. A light type two-layer shallow water model is used for pyroclastic flows lighter than sea water. In this case, light pumice and ash are assumed to be the dominant components of the flow, and they thus travel over the sea surface. Maeno and Imamura [140] applied their model to simulate the tsunamis generated during the 1883 Krakatau eruption. Volumes of 5 to 20 km^3 of pyroclastic flow with densities of 900 to 1500 kg/m^3 and average discharge rates of 10^6 to 10^8 m^3/s were examined. Simulation results suggest that a pyroclastic flow entering the sea with a volume larger than 5 km^3 and an average discharge rate of about 10^7 m^3/s would be the most plausible mechanism for the generation of the tsunami.

Waythomas and Watts [138] evaluated tsunami generation by pyroclastic flow using the event associated with the tsunami deposits of the 3.5 ka eruption of Aniakchak volcano in Alaska. They modelled a pyroclastic flow entering the ocean, characterized by a volume V, maximum velocity U, width w and duration t_o. From these quantities, the characteristic mass flux per unit width into the ocean can be written as $Q = V/(wt_0)$. Considering the local water depth d, a maximum tsunami amplitude h and a characteristic wavelength l_o were estimated using the semi-empirical equations of Walder et al. [154], implemented in the software named TOPICS. Their results suggest that about 25–30% of a volume >50 km^3 (i.e. ~15 km^3) of pyroclastic flow volume had to enter the ocean north of the Aniakchak caldera to generate the observed tsunami deposits.

7.4.3.4. Tsunamis generated by phreatomagmatic explosions

Phreatomagmatic explosions generally occur in shallow water. In these conditions, the wave generation process is relatively simple and well understood, if modelled as a generic underwater explosion (see Fig. 57). Wave generation processes occur when $d_w = W^{1/3}$, where d_w is the depth of the explosion crater and W is the explosion energy in pounds of TNT [155, 156]. The near surface explosion may produce a water crater and even expose the seafloor to the atmosphere. The initial water elevation is assumed to have a crater shape with a watery rim, and an empirical relationship between the explosion energy E and initial wave height η_i is used (e.g. Ref. [151]):

$$\eta_i = 0.024E^{0.24} \tag{15}$$

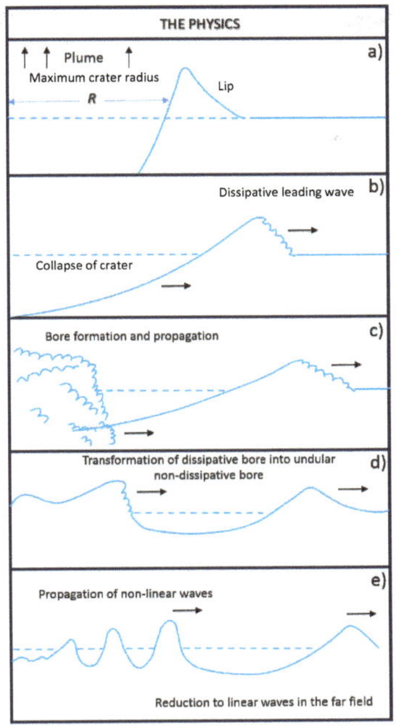

FIG. 57. Schematic representation of shallow water explosion phenomena. Reproduced courtesy of the US Defense Threat Reduction Agency [155].

The crater shape is typically described by an empirical formula (e.g. Ref. [155]):

$$\begin{cases} \eta(r) = \eta_i \left[2\left(\dfrac{r}{R}\right)^2 - 1 \right] \text{ for } r \leq R \\ \eta(r) = 0 \text{ for } r > R \end{cases} \tag{16}$$

where R is the initial water crater radius and r denotes the distance from the source point. Concerning the relation between the initial water source radius and the explosion energy, Le Mehaute and Wang [155] provide an empirical relationship that is valid for shallow water, suggesting a power-law dependence:

$$R \approx 10W^{0.30} \tag{17}$$

where R is given in feet and W in pounds of TNT equivalent (1 ft = 0.3048 m; 1 lb of TNT $\cong 1.9 \times 10^6$ J). In the case when the water reaches the seafloor, the radius of the cavity, R_c, is given by Le Mehaute and Wang [155]:

$$R_c = 4.4W^{0.25} \tag{18}$$

A similar method was also used by Torsvik et al. [151] to analyse a tsunami induced by a phreatomagmatic explosion at Karymsky Lake ($E = 10^{12}$ J), satisfactorily reproducing the wave characteristics. The relationship in Eq. (15) was assumed to be valid even for larger scale explosions by Maeno and Imamura [140], who investigated the possibility of tsunamis generated by a large scale phreatomagmatic explosion during the 1883 Krakatau eruption. In this case, using the empirical relationship between crater size and explosion energy [152], the explosion energy needed to produce a large crater like the Krakatau caldera (2–3 km in radius) can be estimated to be 10^{16} to 10^{17} J, whereas the ratio $d_w/W^{1/3}$ can vary from 0.2 to 0.5. Therefore, a potential largest phreatomagmatic explosion may be associated with a shallow water wave generation process, in which the water, initially expelled upwards and outward, can form a plume and a crater with a watery rim. For this kind of explosion energy, the time required to generate a watery rim is of the order of 1–10 s.

7.4.4. Verification and validation of the tsunami source model

For volcanic landslide triggered tsunamis only, some documented events are available, which can be used for verification and validation. Among these, there is the tsunami produced by the eruption of Augustine Volcano, Alaska, in 1883 [157], which has left recognized traces over a wide area with runups of at least 19 m. The last eruption causing tsunami waves was that of the Karymsky Volcano, Kamchatka, in 1996 [158]. A rock avalanche caused tsunami waves in Karymsky Lake, with runups reaching 30 m that wiped out the surrounding forest [151].

Some modelling examples applied to actual cases (e.g. Refs [127, 149, 151]) are described in Section 7.4.3. So far, because of the low frequency of volcano triggered events, no detailed runup survey data are available, nor are any actual measurements or detailed descriptions of triggering phenomena and tsunami wave propagation and effects. Apart from landslide cases, documentation is available only for a few cases, such as the great 1883 Krakatau eruption (e.g. Refs [136, 159, 160]), for which documentation is still insufficient to achieve a common consensus about this complex event (e.g. Refs [120, 161]). Choi et al. [162] have used the arrival times of the 1883 Krakatoa wave worldwide to validate their model of transoceanic propagation.

7.4.5. Specification of scenario volcanic source

In principle, a single scenario is not sufficient to characterize the volcano related tsunami hazard, since several different source mechanisms are possible. Therefore, a set (or even better, a combination) of potential scenarios that consider different tsunamigenic events should be developed for each of the foreseeable volcanic sources potentially affecting the target site. The starting point is, therefore, the filtering of all capable volcanic edifices within a threshold distance before estimating the ranges of the main controlling parameters according to the type of source mechanism and estimated maximum size of tsunamigenic event.

7.4.6. Size and temporal distribution for probabilistic studies of volcanoes

Probabilistic volcanic hazard assessment can be applied in the site evaluation process [163]. However, the resulting uncertainties can be very large because a sufficient database is rarely available. In principle, it is possible to focus this approach on volcanic tsunamigenic events. Nevertheless, the currently available base of data has only a limited number of entries, only a few of which are sufficiently characterized.

8. TSUNAMI PROPAGATION ANALYSIS

8.1. BASIC CONCEPTS

For assessing the design water level, the numerical calculation method should be adopted that can provide the best estimate of the maximum water rise and drawdown at the nuclear power plant site with consideration of uncertainty.

8.1.1. Selection of the appropriate numerical calculation model

A fundamental framework for applying the governing equations has to be used, along with a numerical scheme that can provide the best estimate of the maximum water rise and drawdown at the nuclear power plant site under adequate conditions of the initial water surface and boundary, taking into consideration uncertainty (see Section 8.5).

8.1.2. Adequate execution of numerical calculation

The computation domain, grid size, time interval, bathymetric and topographic data, coefficients in governing equations and simulation time should be appropriately determined, according to the spatial tsunami shape and geographical features of the seafloor and land areas, along with the source regions and at the nuclear power plant site.

8.2. SELECTION OF A NUMERICAL MODEL

The typical numerical model for tsunami propagation analysis is a two dimensional (2-D) model. The computation area is large, more than several hundred square kilometres at a minimum. The 2-D model is applicable for long wave propagation in large computation areas and can also simulate tsunami runup on land. On the other hand, a three dimensional (3-D) numerical model has recently been developed in the fluid dynamics field. The 3-D model is more appropriate than the 2-D model for tsunami runup with a complex topography and structures. For example, a numerical analysis that combined a 2-D model in the ocean and a 3-D model on land was carried out [164, 165]. In the future, the above-mentioned hybrid model is likely to be commonly used.

8.2.1. Governing equations and numerical schemes of the 2-D model

When the numerical calculation is carried out, factors such as accuracy about the phenomena to be calculated and the necessary computation time should be taken into account, and the appropriate governing equations and numerical schemes should be selected.

8.2.1.1. Governing equations for tsunami evaluation

Since the tsunami has a longer wavelength compared with the water depth, long wave theories in linear or non-linear form with friction and/or dispersion are generally applied. In landslide tsunami cases, where the initial water profiles have a high amplitude and steep slope, non-linear dispersion theory may be necessary. These approaches are described below and are based on 2-D equations that are derived from 3-D governing equations by being integrated in the vertical direction. One of them should be selected by the objective phenomena.

General 2-D equations including all terms are as follows:

$$\frac{\partial \eta}{\partial t} + \frac{\partial M}{\partial x} + \frac{\partial N}{\partial y} = \frac{\partial \xi}{\partial t} \tag{19}$$

$$\begin{aligned} &\frac{\partial M}{\partial t} + \frac{\partial}{\partial x}\left[\frac{M^2}{D}\right] + \frac{\partial}{\partial y}\left[\frac{MN}{D}\right] + gD\frac{\partial \eta}{\partial x} + \frac{gn^2}{D^{7/3}} M\sqrt{M^2+N^2} \\ &\qquad = K_{hf}\left(\frac{\partial^2 M}{\partial x^2} + \frac{\partial^2 M}{\partial y^2}\right) + DISP_x \end{aligned} \tag{20}$$

$$\begin{aligned} &\frac{\partial N}{\partial t} + \frac{\partial}{\partial x}\left[\frac{MN}{D}\right] + \frac{\partial}{\partial y}\left[\frac{N^2}{D}\right] + gD\frac{\partial \eta}{\partial y} + \frac{gn^2}{D^{7/3}} N\sqrt{M^2+N^2} \\ &\qquad = K_{hf}\left(\frac{\partial^2 N}{\partial x^2} + \frac{\partial^2 N}{\partial y^2}\right) + DISP_y \end{aligned} \tag{21}$$

where η is the water surface elevation, M and N are the depth averaged volumetric flux in the (x, y) horizontal directions, respectively, D $(= d + \eta)$ is the total water depth, d is the water depth, ξ is the sea-bottom deformation by crustal movement, n is Manning's roughness coefficient, K_{hf} is the horizontal diffusion coefficient, $DISP_x$ and $DISP_y$ are dispersion terms, g is gravity acceleration and t is time. The above three equations are solved numerically for the unknown values of M, N and η.

(a) Linear long wave theory

Linear long wave theory is applied under the condition that the ratio of the wave height to the water depth is sufficiently small. The equations of motion comprise an unsteady term and a pressure term (hydrostatic distribution). This theory is sufficient to simulate the propagation of tsunami waves in the deep ocean.

$$\frac{\partial M}{\partial t} + gd\frac{\partial \eta}{\partial x} = 0 \tag{22}$$

$$\frac{\partial N}{\partial t} + gd\frac{\partial \eta}{\partial y} = 0 \tag{23}$$

(b) Non-linear long wave theory (shallow water wave theory)

This is applied under the condition that the ratio of the wave height to the water depth is not small (the non-linearity cannot be ignored, especially in the shallow region). The equations of motion comprise an unsteady term, a pressure term (hydrostatic distribution) and an advection term; with these terms, the steepening of the wavefront in shallow water can be considered. In general, friction becomes important in the shallow zone and on land, and the bottom friction term must be used. The horizontal eddy viscosity term may be considered, if necessary.

$$\frac{\partial M}{\partial t} + \frac{\partial}{\partial x}\left[\frac{M^2}{D}\right] + \frac{\partial}{\partial y}\left[\frac{MN}{D}\right] + gD\frac{\partial \eta}{\partial x} + \frac{gn^2}{D^{7/3}}M\sqrt{M^2+N^2} = K_{hf}\left(\frac{\partial^2 M}{\partial x^2} + \frac{\partial^2 M}{\partial y^2}\right) \tag{24}$$

$$\frac{\partial N}{\partial t} + \frac{\partial}{\partial x}\left[\frac{MN}{D}\right] + \frac{\partial}{\partial y}\left[\frac{N^2}{D}\right] + gD\frac{\partial \eta}{\partial y} + \frac{gn^2}{D^{7/3}}N\sqrt{M^2+N^2} = K_{hf}\left(\frac{\partial^2 N}{\partial x^2} + \frac{\partial^2 N}{\partial y^2}\right) \tag{25}$$

(c) Dispersive wave theory

This is applied in conditions in which the curvature of the tsunami wave increases with propagation, the vertical acceleration of the water particles cannot be ignored, and wave dispersion appears. The dispersion theory is classified into the linear dispersive wave theory and non-linear dispersive wave theory, depending on the theories to which the dispersion terms are added.

Dispersion is important during long distance propagation and in the shallow region. In deep water, the linear dispersion theory may be applicable. In shallow regions, the non-linear dispersion theory should be used. Non-linear dispersion must be considered for simulating short, steep and high amplitude waveforms, such as with landslide generated tsunamis. For near field tsunamis, wave breaking occurs before or after runup if a tsunami is accompanied by soliton fission. It was proposed that the non-linear dispersive wave theory is better suited to the entire numerical area, which includes the tsunami source and the coastal area, since it can evaluate the tsunami wave shape more precisely than the non-linear wave theory without reference to soliton fission. This is because the dispersive term has the effect of suppressing the leaning of the wavefront in the deep and shallow sea areas. In the future, the dispersive wave theory is expected to facilitate the development of a more precise and practical numerical model than the non-linear long wave theory by including a damping term due to wave breaking. Some tsunami numerical models might have numerical dispersion, which appropriately mimics frequency dispersion with an optimized grid resolution, such as the Method of Splitting Tsunami (MOST) numerical model [166–168].

8.2.1.2. Governing equations and numerical scheme for near field tsunami propagation

For nearshore tsunami propagation, where the water depth is shallower than 200 m, the governing equations of the non-linear wave theory should be selected [169]. In such a case, an explicit finite difference scheme with a staggered leapfrog method is generally adopted because the analysis method of the numerical error caused by the finite difference scheme is well established [170].

8.2.1.3. Governing equations and numerical schemes for far field propagation of tsunamis

In the case of the far field propagation of a tsunami, the linear theory can be applied because the wave height is small when compared with the water depth. However, when the initial tsunami profile contains a wide range of wavelength components, the wave velocity differs slightly for each wavelength in the deep water; further, since it propagates for a long time, the delay of the shorter wave becomes larger. Therefore, to reproduce this effect, it becomes necessary to apply governing equations that include the dispersion term.

Furthermore, for distant tsunamis, the Coriolis force must be considered in the equations of motion. Also, since the effects of the spherical Earth cannot be ignored, a spherical coordinate system must be adopted. For a numerical scheme, the alignment of variables is usually performed by a staggered leapfrog scheme, and the explicit difference method is adopted for the equation of continuity. An implicit scheme is generally adopted for the equations of motion.

8.2.2. Initial conditions

The initial conditions of tsunami numerical modelling are the vertical elevations of water surface and discharge fluxes in lateral directions throughout the computational domain at the time of tsunami initiation. These initial conditions for seismic tsunamis are ocean bottom deformation computed from fault parameters (see Section 7.2.3). The initial conditions given at the ocean bottom must be converted to those on the water surface, considering the wavelength and water depth (see Section 7.2.3.3).

The initial conditions related to non-seismic mechanisms may cover the time change of the vertical displacements of water level and lateral fluxes regarding the triggering mechanisms. Estimating the lateral fluxes and even the water surface elevations is rather complex, since more investigations are needed to develop verified methods for the case of non-seismic triggering mechanisms. Various methods may be used to estimate the initial conditions of tsunami sources due to submarine and/or subaerial landslides (see Section 7.3) and/or volcanic activity (see Section 7.4).

8.2.3. Boundary conditions

When carrying out the numerical calculations, the following boundary conditions, associated with the computational region, submarine and coastal topography, structures, and so on, should be properly applied: (1) offshore boundary conditions, (2) onshore boundary conditions and (3) overflow boundary conditions.

8.2.3.1. Offshore boundary conditions

Since the computational region is finitely determined, open boundaries are artificially provided on the offshore side(s) of the model domain. Appropriate boundary conditions need to be applied so that the behaviour of the tsunami is free from artificial reflection from the boundaries.

(a) Open boundary condition

An open boundary condition is necessary for waves propagating toward a boundary to radiate to the outer end of the boundary of the numerical domain with negligible numerical reflective waves. The typical methods for expressing the discharge flux using the radiative wave conditions [88, 170] can be adopted. The method that uses free transmission conditions was proposed for situations in which this condition is not satisfied; this method was based on the method of characteristics [170].

(b) Boundary condition for incident tsunami

For evaluation of a distant tsunami, the tsunami propagation in the large extent of the sea has to be calculated by the linear dispersive wave theory, expressed by the spherical coordinates. If the coordinate system of the domain near the site is presented by rectangular coordinates, it is difficult to carry out the numerical simulation considering the interaction of water level and flux between both the outer and the near site domains. In such a case, the time series of water level changes and flux, given by numerical results in the outer domain, may be used for the offshore boundary condition of the near site domain.

Alternatively, the time series of the tsunami predicted from the actual data at the near field can be used as an input for the offshore boundary condition.

8.2.3.2. Onshore boundary conditions

The boundary conditions at the leading edge of sea water should be applied by the following conditions.

(a) Reflective boundary condition

When the tsunami runup near the shore is not considered, a vertical wall with an infinite height should be located on the coastline, and the discharge flux in the direction perpendicular to the coastline should be assumed to be zero. In other words, a complete reflection condition should be applied.

(b) Boundary conditions at the runup front

The tsunami inundation or drawdown has to be simulated in tsunami propagating analysis to estimate the impact of a tsunami near the shore. Previous research results and illustrated typical methods on the numerical modelling of tsunami inundation are summarized in Fig. 58 [171]. Two procedures are necessary

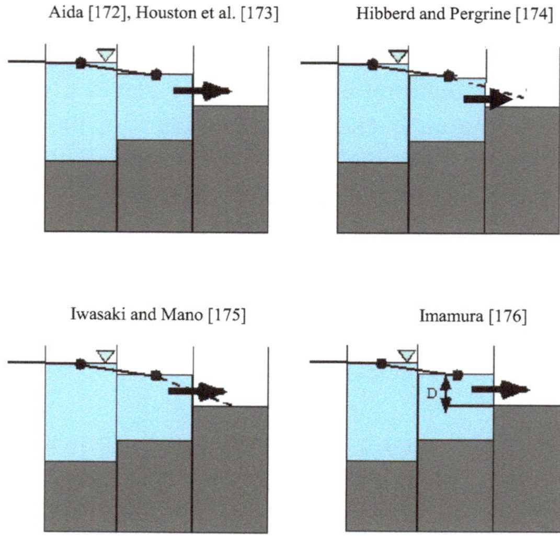

FIG. 58. Some examples of the front of a wave. Several methods to calculate the discharge at the estimated wavefront are presented in Ref. [171]. Reproduced courtesy of Harvard University Press.

as a minimum: the first is searching the wavefront, and the second is setting the discharge at the wavefront in a computational domain. Several methods are introduced to estimate the inflow discharge flux in inundating from a wet grid to a neighbouring dry grid (see Fig. 58).

8.2.3.3. Overflow boundary conditions

The boundary conditions for the case in which a tsunami flows over a breakwater, sea dyke, seawall revetment or other structure can be applied according to the following conditions:

(a) When the structure is modelled as a part of the topography.
  — In this case, the boundary conditions at the runup front described in the previous Section can be applied to the moving boundary conditions in which the tsunami flows over the breakwater or other structures.

(b) When the structures are modelled by the boundaries between grids.
  — When breakwaters or other structures exist in the computational region and the water level exceeds the crest elevation, the discharge that flows over the structure can be estimated using the relevant formulas by the overflow conditions.

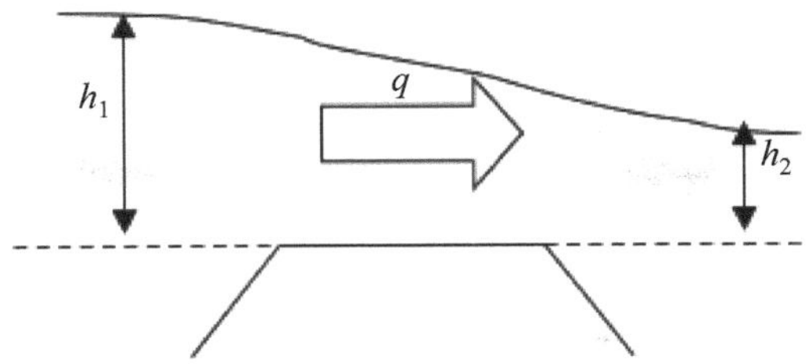

FIG. 59. Definition of q, h_1, and h_2 in Hom-ma formulas for boundary conditions of overflow at the breakwater. Reproduced courtesy of Japan Society of Civil Engineers [91].

Hom-ma formulas [170, 177] can be used to determine overflows and can be summarized as follows (see Fig. 59):

For complete and incomplete overflows,

$$q = \mu h_2 \sqrt{2gh_1},\ h_2 \leq \frac{2}{3} h_1 \tag{26}$$

For submerged overflows,

$$q = \mu' h_2 \sqrt{2g(h_1 - h_2)},\ h_2 > \frac{2}{3} h_1 \tag{27}$$

where h_1 and h_2 are the water heights in front of and behind the structure, respectively, measured from the surface equal to the top of the structure; $\mu = 0.35$; $\mu' = 2.6\mu$; and g is the gravity acceleration.

If the water does not flow over the breakwater and sea dyke, the complete reflection condition setting of the vertical wall is assumed, and the discharge flux in the direction perpendicular to the structures is assumed to be zero.

8.3. NUMERICAL CONDITIONS

8.3.1. Specifying the numerical domain and spatial grid size

The computational region, including the tsunami source, should be set up such that refraction (including lens effects), reflection (including multiple reflections), diffraction, seiches, trapping phenomena (around islands or continental shelf), runup, and so on, all of which have considerable effects on the maximum water rise and drawdown at the site, can be reproduced with high accuracy.

In the tsunami calculation, the method of connecting computational regions with varying mesh sizes by the tsunami profile and topographical conditions is used from this viewpoint. The calculation of the tsunami is performed simultaneously within the connecting computational regions. In other words, the wavelength of a tsunami in the open sea is on the order of several tens of kilometres to several hundred kilometres, and it decreases as the water depth decreases. It is necessary to change to successively finer mesh sizes to meet the above conditions. Also, since nearshore bathymetry and land topography are usually complex, the spatial mesh size must be properly set up by the characteristic topography of the seashore concerned or the scale of the artificial structures, as well as the spatial scale of the tsunami profile.

8.3.2. Specifying the time interval

The time interval Δt is set to satisfy the Courant–Friedrichs–Lewy (CFL) condition in terms of the calculation stability. The CFL condition for the horizontal 2-D numerical calculation is as follows:

$$\Delta t \leq \frac{\Delta x}{\sqrt{2gd_{\max}}} \tag{28}$$

where Δx is the mesh size, Δt is the time interval, d_{max} is the maximum water depth and g is the gravity acceleration. Multiple calculation regions with different grid sizes Δx are generally adopted for numerical calculations. The time intervals Δt for all calculation regions are most often set to be the same and constant. In that case, all the time intervals Δt must satisfy the CFL condition. Finally, the minimum Δt in all regions is adopted as the time interval.

However, when the calculation is performed in practice, numerical errors and non-linearity of the phenomena interfere. Hence, Δt must often be set at a smaller value as compared with $\Delta x / \sqrt{2gd_{\max}}$, an allowance being provided. In particular, in the case of calculations with a strong current due to tsunami runup, the flow velocity becomes greater than the wave celerity of the tsunami, $\sqrt{2gd_{\max}}$; this might result in a divergence of the calculation.

8.3.3. Topography data and bathymetry data

In principle, topographic data used for the calculation should be prepared using the latest information on the design tsunami and structures at the site (see Section 6.4).

8.3.4. Specific parameters

Specific parameters used for the numerical calculations should be appropriately chosen by referring to previous studies.

8.3.4.1. Coefficients related to the friction term

The coefficients of the friction term can be chosen in accordance with Table 4.

TABLE 4. COEFFICIENTS ASSIGNED TO THE FRICTION TERM
(reproduced courtesy of Japan Society of Civil Engineers [91])

Name of coefficient	Values reported in previous studies	Values used frequently for assessment of water level of design tsunamis of nuclear power plants
Manning's coefficient of roughness n ($m^{-1/3}/s$)	Iwasaki and Mano [175]: 0.03 for the sea area Goto and Sato [178]: 0.025 for the sea area Kotani et al. [179]: Inundation area: High density residential district: 0.08 Medium density residential district: 0.06 Low density residential district: 0.04 Forest area: 0.03 Agricultural area: 0.02	Sea area: 0.03 Inundation area: 0.03
Friction coefficient k_b	Tanaka [180]: Deep-sea area: 0.0026 Shallow sea area: 0.005–0.01 Inundation area: 0.01–0.5	Deep-sea area (usually deeper than 15 m): 0.0026 Shallow sea area (usually shallower than 15 m): 0.00637 Inundation area: 0.01

However, when the friction coefficient is varied by the water depth, the current velocity field may yield unnatural results if the variation is discontinuous. Hence, it is recommended to set the friction coefficient such that a smooth variation is ensured.

8.3.4.2. Coefficient of eddy viscosity

In the past, a coefficient of eddy viscosity smaller than 10 m^2/s (10^5 cm^2/s) was used. If the water level change is subject to assessment, 10 m^2/s (10^5 cm^2/s) can be used as the actual maximum value of the coefficient of eddy viscosity. The absolute percentage error in wave amplitude, compared with the use of a coefficient of eddy viscosity of zero, is approximately 5% or less. The 10 m^2/s criteria may need to be further evaluated, depending on the site characteristics.

8.3.5. Simulation duration

The simulation duration should be appropriately set after taking into account the tsunami characteristics, topographical conditions and so on.

A tsunami does not always cause the maximum water rise or drawdown with the first wave [181]. Moreover, the times of maximum rise and drawdown vary by the time series of water level changes at the wave source and morphological conditions in the region of the sites.

When the resonant oscillations inside the bay or seiche are excited, or the reflection waves of the first wave from the opposite bank and a subsequent tsunami are superimposed, either a maximum water rise or the maximum water drawdown may occur. Ocean ridges could act as waveguides, or waves trapped on the continental shelf could result in higher tsunami heights, and a larger duration of tsunami propagation might be necessary to capture these phenomena. For this reason, it is important to select a simulation duration that is suitable for analysing these phenomena.

8.4. VERIFICATION AND VALIDATION OF NUMERICAL MODEL

The numerical model used for tsunami propagation should undergo verification and validation through benchmark problems [182]. Verification is the process of ensuring that the model accurately solves the parent equations of motion, and validation is the process of ensuring that the model represents geophysical reality. These definitions are similar to those used in other guidelines provided by the IAEA but are different from — exactly opposite to — those adopted by NOAA [182, 183].

A numerical calculation system is validated by applying numerical calculations to (a) analytical problems and (b) experimental problems and is verified by applying them to historical tsunamis [183–185].

8.5. UNCERTAINTIES IN NUMERICAL MODELS

As tsunamis are natural phenomena, their variable and uncertain aspects should be considered. Because the runups and inundation heights calculated by the numerical model mentioned above include uncertainties associated with (i) data (source parameters, bathymetry, topography; see Section 6), (ii) the physics adopted to describe the source (e.g. the fault model for an earthquake) and propagation (e.g. the appropriateness of shallow water equations, non-linear effects; see Sections 7 and 8), and (iii) numerical calculation (grid size, truncation error, finite difference; see Section 8), the observed values distribute around the calculated values.

If the distribution of tsunami runups around the median value is assumed to be log-normal on the distribution of the ratios between observed and calculated values, the uncertainty in the numerical model is calculated as Aida's κ [90], considering all uncertainties that cannot be divided (see Section 5).

In the case of the probabilistic tsunami hazard assessment (see Section 9), it is necessary to take into consideration the distribution of the tsunami runup in the time domain at the site. However, its distribution in the time domain is not obtained, so the ergodic hypothesis is applied. The ergodic process is a random process in which the distribution of a random variable in the space domain is the same as the distribution of that same random variable in the time domain at a single point when sampled as a function of time. Therefore, in the probabilistic tsunami hazard assessment, the variance parameter β in the time domain, logarithm standard deviation, is related to κ, an error factor often referred to as Aida's index, as $\beta = ln(\kappa)$ [186].

9. TSUNAMI HAZARD ASSESSMENT OF COASTAL IMPACTS

9.1. DETERMINISTIC APPROACH

9.1.1. Concept of the deterministic approach

The design basis tsunami should provide an adequate level of safety to nuclear power facilities in consideration of all historical and possible future tsunamis. In the deterministic approach, this is achieved by estimating the potential tsunami source that would produce the probable maximum tsunami height at the site, on the basis of the historical and geophysical data, whereas the probabilistic approach explicitly considers the inherent uncertainty associated with the tsunami height at the site. Because tsunamis are natural phenomena, uncertainties and errors in many important parameters are unavoidable, hence they need to be considered in the tsunami hazard assessment. A systematically conducted sensitivity analysis or parametric study should be used to support the evaluation of the significance of the contributions of the various input parameters in the model. The design basis tsunami should have the most significant impact on the nuclear power plant site in terms of probable maximum and minimum water levels. Results in the range of observed or recorded historical data should be validated. The overall procedure by the deterministic approach proposed by JSCE [91] is shown in Fig. 60 and described below.

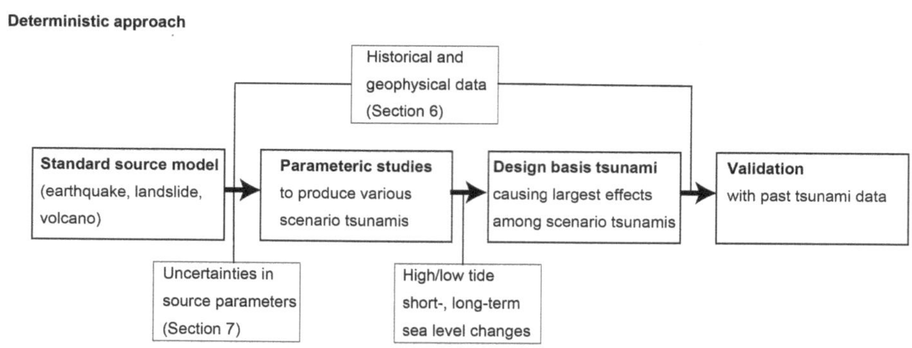

FIG. 60. Flowchart for the assessment process for the design basis tsunami in the deterministic approach [91].

9.1.1.1. *Standard source model*

A standard source model is selected from a literature survey or constructed with consideration paid to the historical and geophysical data (see Section 6). The source parameters of earthquake, landslide or volcanic origin of the standard source model (see Section 7), as well as the uncertainties of the parameters, are specified.

9.1.1.2. *Parametric studies*

To account for the uncertainties regarding a tsunami source, a large number of numerical calculations are carried out under various conditions within a specific range of parameters. The range is determined by the estimated uncertainties. This is referred to as a 'parametric study'. The source models considered in the parametric study are termed 'scenario tsunamis'. The effects on the nuclear power plant site — not only the maximum water level but also the maximum drawdown at the site — are calculated for each scenario source.

9.1.1.3. *Design basis tsunami*

Among the various scenario tsunami sources for each area, the one causing the largest effect, either maximum water rise or drawdown, at the nuclear power plant site is selected as the 'design basis tsunami'. The scenario for the maximum water rise tsunami may be different from the scenario causing the maximum drawdown. The design water level is defined as the sum of the 'design basis tsunami' and an appropriate tidal condition and long term sea level change.

9.1.1.4. *Validation of the design basis tsunami*

The design basis tsunami must be validated using the historical data to confirm that the design basis tsunami height exceeds all the recorded and calculated historical tsunami heights at the nuclear power plant site. In the vicinity of the nuclear power plant, the envelope of the scenario tsunami heights must exceed all recorded and calculated historical tsunami heights.

9.1.2. Example for earthquake sources

For earthquake sources, the location, depth, dimension (length, width and average amount of slip) and geometry (strike, dip and slip angles) of faults must be specified (see Section 7). The dynamic parameters such as rise time or rupture velocity may affect the tsunami generation. Non-uniform distribution of slip and multiple ruptures of segments may also need to be considered for large local earthquakes [187] (see Section 7.2.2.3). In a parametric study, these input parameters are varied within their range of estimated uncertainties. Uncertainties associated with the estimated source parameters have been discussed in detail in Section 7.

In the numerical simulation, the governing equations, boundary conditions, initial conditions, grid division, modelling of bathymetric and topographic data and observed runup heights for historical scenario verification may also include uncertainties. However, it is rather difficult to estimate those uncertainties quantitatively and to deal with them one by one. Only uncertainties concerning the tsunami source are dealt with in this parametric study because they can significantly influence tsunami assessment.

A concept of a parametric study of the design basis for earthquake generated tsunamis developed by JSCE [91] is shown in Fig. 61. The upper part of the figure shows fault models for scenario earthquakes. Each rectangle with a dashed line represents a fault model. In the lower part of the figure, each curve represents a scenario tsunami, which is calculated on the basis of each fault model. Similar studies can be developed for other tsunami sources.

Other water level changes such as tides and short and long term sea level change need to be considered, as appropriate. JSCE [91] has recommended that a high or low tide datum (for maximum water levels and largest drawdown, respectively) be used for the numerical simulations to include non-linear hydrodynamic effects near and onshore.

The nuclear power plant design has to have sufficient margin above the design basis tsunami (i.e. all historical and possible tsunamis at the site) to ensure the safety of nuclear power plants sited on the coast (see Fig. 61). It has to be noted that the tsunami sources that give rise to the maximum water levels and those that result in the largest drawdown are sometimes different.

Many of the tsunami source parameters are related through scaling relations. The overall size of the 'standard source model' is determined. The maximum source size is estimated on the basis of the frequency–size relation or geophysical characteristics of the source (Section 7.2.6). Some studies have provided estimates on the earthquake frequency–size distribution and maximum magnitude for global subduction zones [92, 96, 97].

Based on the statistical analysis of the results from parametric studies and the approach to estimation of the upper bound values of source size, it is possible to select design basis tsunami heights corresponding to the maximum and minimum wave heights at the site. A distribution expressing the total uncertainty in wave height from the parametric study may be referenced in probabilistic analysis (see Section 9.2). This can be either an empirical distribution function

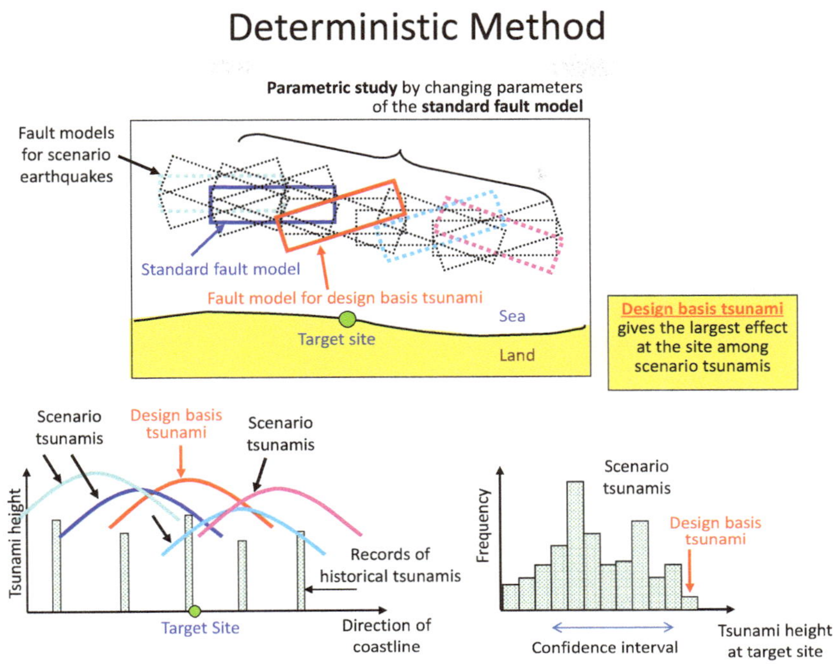

FIG. 61. Concept of setting up source fault and parametric studies following the JSCE method [91].

(e.g. derived from the histogram, lower right of Fig. 61) or a fitted analytic distribution function.

When the predominant period of the tsunami and the natural period of free oscillation for the harbour or the intake passage is equal, the water rise and fall may be amplified by seiche activity (Section 3). The effect of resonance needs to be investigated by numerical simulations.

Other associated phenomena, such as the movement of sand sediment, inundation from an adjacent river and coseismic uplift or subsidence from nearby capable faults, have to be evaluated on the basis of specific site conditions (Section 10).

9.2. PROBABILISTIC APPROACH

9.2.1. Concept of PTHA

In contrast to deterministic analysis, PTHA determines the level of hazard for a specific design probability or mean return period. The hazard from a large number of possible sources is aggregated to develop a tsunami hazard curve that plots runup or another hazard variable along the horizontal axis and the frequency or probability of exceedance along the vertical axis (see Fig. 62). Furthermore, to include many different types of tsunami generators, multiple sources of uncertainty related to source parameters and numerical models of tsunamis are considered in PTHA. Two kinds of uncertainty, aleatory and epistemic, are distinguished in the PTHA (see Section 5.1). A single hazard curve is obtained by integration over the aleatory uncertainties and all possible sources. A large number of hazard curves are obtained for different branches of the logic tree representing epistemic uncertainty, as indicated by Fig. 62 (left). The final hazard curve at the nuclear power plant site is represented by the mean, median or other specified fractiles of all-hazard curves (see Fig. 62 (right)).

9.2.2. Outline of PTHA procedure for all sources

A diagram showing the general procedure for PTHA is shown in Fig. 63. This procedure is valid for all sources of tsunamis.

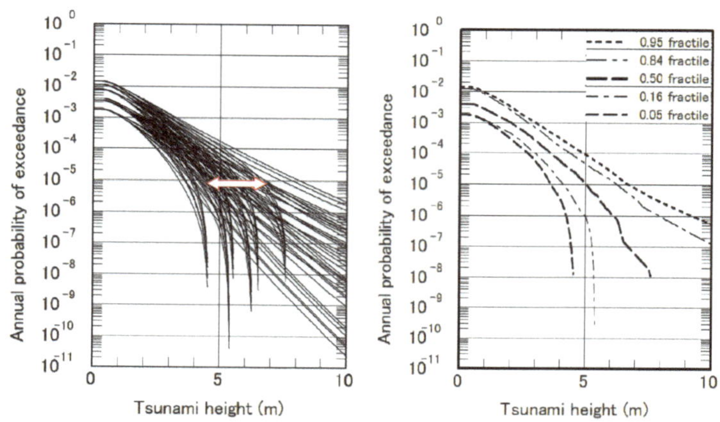

FIG. 62. Example of tsunami hazard curve. Left: all-hazard curves produced from a logic tree. Right: the final tsunami hazard curve determined as a fractile of all-hazard curves. Five different fractiles are displayed. Reproduced courtesy of Springer Science+Business Media [186].

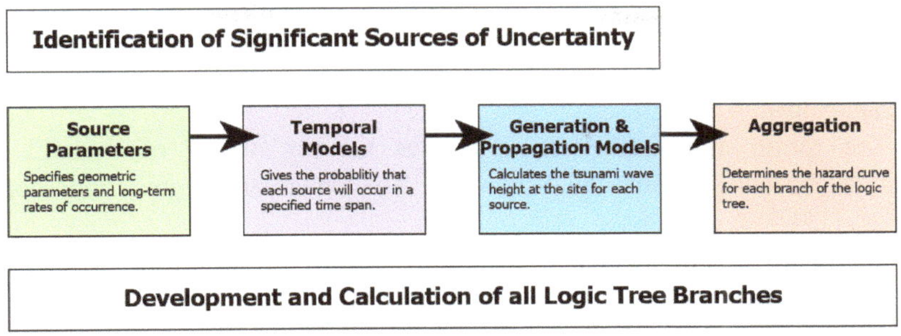

FIG. 63. Schematic of PTHA procedure. Reproduced courtesy of The Oceanography Society [188].

Before the analysis begins, significant sources of uncertainty should be identified (top row of Fig. 63). These can be classified as either epistemic or aleatory. Consideration should be given to the expected effect each uncertainty has on the hazard curve at the design probability of interest. Typically, a panel of experts is assembled to make this determination; the technical integrator/facilitator approach, for example [189, 190]. The logic tree represents alternative hypotheses and interpretations [186], and the weights for each branch can be determined subjectively by the experts or objectively sampled, for example, from a uniform or normal distribution. Ideally, the branches of a logic tree should be mutually exclusive and collectively exhaustive. At each node of the logic tree, the weights must sum to 1.

The sequential procedure for each branch of the logic tree is given in the middle row of Fig. 63. First, the distribution of source parameters is determined. Typically, a parameter representing the overall size (e.g. seismic moment for earthquakes, volume for landslides) and its distribution is defined. Other source parameters are scaled concerning the size parameter. Different size distributions and scaling relationships can be considered to represent epistemic uncertainty (i.e. different logic tree branches). Importantly for PTHA, a relationship between the tsunami source size and the long term rate of occurrence must be established.

Second, a temporal model for the occurrence of the sources is chosen. Often, a Poisson process is assumed, with an exponential distribution of interevent or recurrence times. However, alternative probability models can be considered in the logic tree framework, such as quasiperiodic and temporally clustered models.

Third, for each source used in the PTHA analysis, a tsunami generation, propagation and runup/inundation model is used to determine the wave height or other hazard variable at the nuclear power plant site. Tsunami generation and propagation models are described in Sections 7 and 8. Again, if uncertainty remains about which model best describes a particular phase of the tsunami process, alternative models can be included as epistemic uncertainty. The performance of a particular model in reproducing tsunami observations can be considered a source of uncertainty, often represented by a truncated log-normal distribution [186] as described below.

The fourth and final step aggregates the probability of tsunami wave height exceedance from all sources. The general aggregation formula (e.g. Ref. [191]) is given by:

$$P(R \geq R_0|T) = 1 - \Pi_{j=1}^{N}\left[1 - P_j\left(R \geq R_0|T\right)\right] \tag{29}$$

where $P_j(R \geq R_0|T)$ is the probability of wave height exceedance over time T from scenario j. If a Poisson process is assumed for each source (Step 2), then the probability is expressed by an exponential distribution:

$$P\left(R > R_0, T\right) = 1 - e^{-\lambda T} \tag{30}$$

The total rate of tsunami hazard exceedance *(λ)* is the sum of the exceedance rates associated with each source. The annualized probability (T = 1 year) is approximately equal to the rate if the rate is small. That is,

$$P\left(R > R_0, T = 1\right) = 1 - e^{-\lambda} \approx \lambda \tag{31}$$

Aleatory uncertainty is included by integrating the probability distribution of tsunami wave height associated with the uncertainty [192, 193]:

$$\lambda\left(R \geq R_0\right) = \Sigma_{type=i}\Sigma_{zone=j}v_{ij}\int P\left(R > R_0\middle|\Psi_{ij}\right)f(\Psi_{ij})d\Psi_{ij} \tag{32}$$

where v_{ij} is the source rate, $f(\Psi_{ij})$ is the probability density function for source parameters Ψ_{ij}, such as a power-law type distribution for seismic moment encompassed by the GR relation, and $P(R > R_0|\Psi_{ij})$ is expressed as a cumulative probability distribution of runup or wave height exceedance given aleatory uncertainty in the source and model parameters. $P(R > R_0|\Psi_{ij})$ is determined from the tsunami generation and propagation models in step three and can often be fit by an analytic distribution, such as normal or log-normal distributions. For the latter, which is common in tsunami analysis, the probability density function (p) corresponding to the cumulative distribution (P) is given by:

$$p\left(R \middle| \Psi_{ij}\right) = \frac{1}{\beta R\sqrt{2\pi}} \exp\left[-\frac{1}{2}\left(\frac{\ln R - \mu}{\beta}\right)^2\right] \tag{33}$$

where μ and β are the mean and standard deviation of the natural logarithm of tsunami heights (see Fig. 64). If the computed water height is assumed to be the mean height of the log-normal distribution, the natural logarithm of computed height can be used for μ and the natural logarithm of geometric standard deviation κ (see Section 7) can be used for β, i.e. $\ln \kappa = \beta$.

The rates are summed over all types of tsunami generators (i index), such as earthquakes, landslides and so on, and over all zones for each source type (j index). An example of a zonation scheme is the Flinn–Engdahl zones for earthquakes [194]. Combining all sources and their aleatory uncertainty produces a tsunami hazard curve at the nuclear power plant site. The same procedure is used to produce a hazard curve for each logic tree branch representing the range of epistemic uncertainty.

Throughout this process, the entire logic tree is constructed, keeping track of each branch (bottom row of Fig. 63). Careful attention is paid to making sure the weight at each node of the logic tree sums to 1. After all-hazard curves for each branch of the logic tree are estimated, a mean, median or some other fractile is chosen as the final hazard curve for the nuclear power plant site.

Further analysis may include validation and deaggregation of the results. Empirical exceedance rates can be determined from historical tsunami catalogue data and can be used to validate the PTHA results. If the empirical rates are significantly higher than the PTHA results for a particular site, the assumptions and

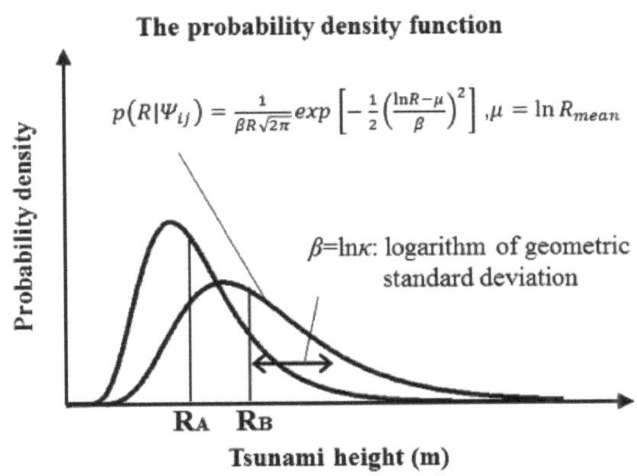

FIG. 64. Log-normal distribution of tsunami height (R_A, R_B: Computed tsunami heights for model A and B).

calculations used in the PTHA procedure should be re-examined. Deaggregation [195, 196] of the PTHA results identifies the dominant source or sources in the PTHA analysis for a given design probability of mean return time. Deaggregation calculates the relative contribution of source locations and parameters to the total hazard. It is useful for identifying potential source scenarios for use in a detailed deterministic analysis and for computing wave time histories (see Section 10.2).

9.2.3. PTHA procedures for earthquakes

A specific example of the PTHA procedure that is listed in Section 9.2.2 and shown in Fig. 63 is given below for earthquake sources. Sources of aleatory and epistemic uncertainty are identified throughout this example. Results from this analysis would be combined with results using other generators of tsunamis, such as submarine landslides, to yield a comprehensive tsunami hazard curve.

9.2.3.1. Step 1: Source parameters

(a) Determination of the tsunami source zones

The source zones for seismogenic tsunamis (i.e. earthquake rupture zones) are determined with respect to tsunami impact at the nuclear power plant site. The source zones are classified on the basis of present knowledge and information from historical earthquakes and tsunamis. In addition to historical earthquake information from literature research, palaeoseismic research, such as the analysis of traces of liquefaction and tsunami deposits, should be considered.

Different hypotheses of earthquake occurrence can be included as different branches of a PTHA logic tree. For example, under a segmentation model, multisegment earthquake ruptures may occur along fault zones (see Section 7). An alternative hypothesis to the segmentation model is that earthquakes are randomly distributed along a particular fault zone without regard to stationary segments. In both cases, only large earthquakes are typically considered for PTHA, since small or medium earthquakes less than M_w 6.5 will not generate large tsunamis. Large distant earthquakes with the potential to generate significant tsunamis at the nuclear power plant site can be considered using similar zonation schemes.

(b) Setting the earthquake magnitude distribution

The moment magnitude M_w of an earthquake hypothesized in the tsunami source zone cannot be specified as only one value. The influence of the specific distribution used for M_w on tsunami generation and the PTHA hazard curve is significant. One hypothesis is the characteristic hypothesis (see Section 7.2.6).

Uncertainty of the characteristic M_w can also be included in PTHA. Multisegment earthquakes are associated with larger values of M_w.

A second hypothesis is that earthquakes occur with M_w distributed according to the GR law, parameterized by an activity value (*a* value, i.e. the source rate parameter v_{ij} in Fig. 63) and a power-law exponent (*b* value). Some examples of *a* and *b* values for particular seismic zones are given (see Section 7.2.6).

(c) Design of the fault model

The fault model for tsunami source zones is designed on the basis of the results of literature research, seismicity analysis (3-D locations of past earthquakes) and geophysical surveys. The primary fault zone parameters are strike and dip (and their variations). It is also possible to refer to past fault models developed by government and scientific organizations, such as the Global Earthquake Model Active Fault database.[1]

(d) Setting other source parameters

Other earthquake source parameters, such as mean slip and rupture dimensions (if not specified by the segmentation or characteristic model), can be scaled in terms of M_w. Uncertainty in the scaling coefficients can be included as uncertainty concerning their effect on tsunami wave heights at the site. Parametric studies, such as those described in Section 9.1, can be used to determine the aleatory probability distribution of tsunami wave height. Some earthquake source parameters such as shear modulus and slip (rake) angle, do not scale with M_w. A review of research literature can be used to establish these parameters. Moreover, the effects of non-uniform slip [187] need to be considered for local sources, as mentioned in Section 9.1.2.

9.2.3.2. *Step 2: Probability model*

Different probability models for earthquake occurrence can be considered in the logic tree framework. A stationary Poisson model in which earthquake recurrence times follow an exponential distribution is typically associated with the GR magnitude distribution. In this case, the probability of occurrence is independent of the time since the previous seismic event. In contrast, a quasiperiodic time dependent model, such as one in which the recurrence times follow a Brownian passage time distribution [101], is typically associated with the characteristic earthquake model. The recurrence times of earthquakes

[1] https://www.globalquakemodel.org/product/active-faults-database

associated with a Brownian passage time distribution have a mean value of μ and a coefficient of variation of α, which are specified as parameters of the distribution. Other quasiperiodic models include those that follow log-normal and Weibull probability distributions. The probability of a multisegment rupture may be estimated using historical earthquake data.

9.2.3.3. *Step 3: Generation and propagation models*

Different models for simulating the tsunami generation process for earthquakes and tsunami propagation modelling are described in Sections 7 and 8, respectively. Because of the large number of calculations associated with PTHA, computationally efficient methods of numerical modelling should be explored. Random model errors in reproducing tsunami observations can be considered a source of uncertainty, often represented by a truncated log-normal distribution [186].

9.2.3.4. *Step 4: Aggregation*

Aggregation of tsunami wave heights using different source parameters, aleatory uncertainty and earthquake probability models to calculate a tsunami hazard curve is described in Section 9.2.2. An apparent Poisson rate (λ) can be calculated for the time dependent probability models, such as the Brownian passage time distribution, so that the aggregation in Eq. (32) can be used [197]. An example logic tree is shown in Fig. 65 and includes epistemic uncertainties associated with the analysis. A hazard curve is computed for each branch of the logic tree, and the selected fractile of all-hazard curves is used to represent the hazard for seismogenic tsunamis at the nuclear power plant site.

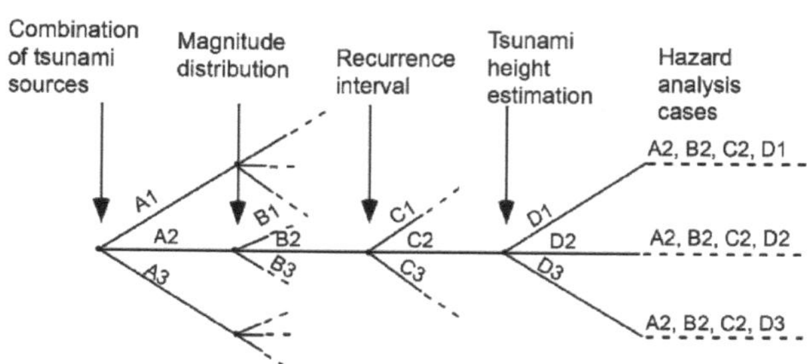

FIG. 65. Example logic tree for seismogenic tsunamis. Reproduced courtesy of Springer Science+Business Media [186].

9.2.4. Example of PTHA studies

Some examples of PTHA studies are summarized in Table 5.

TABLE 5. EXAMPLES OF PTHA STUDIES

Types of study	References
Use of PTHA logic tree for Japan	Annaka et al. [186], Atomic Energy Society of Japan [198]
Incorporating multiple sources of uncertainty into PTHA for California, USA	Thio et al. [199]
Construction of probabilistic inundation map for Seaside, Oregon, USA	Tsunami Pilot Study Working Group [200], González et al. [201]
Regional probabilistic study, incorporating historical data for the Caribbean region	Parsons et al. [202]

10. TSUNAMI LOAD EFFECTS TO NUCLEAR FACILITIES

10.1. GENERAL

The previous section has described procedures to estimate tsunami heights on the basis of the deterministic and probabilistic approaches. However, as discussed in Section 4.1, not only are tsunami effects at nuclear power plant sites related to maximum rise in water level or runup, but they also highly depend on hydrodynamic parameters such as maximum current velocity, flow depth, hydrodynamic forces, maximum drawdown in water level, duration of inundation and of drawdown. Deposition, erosion and scouring due to the tsunami flow around the sites or collision of debris with structures also affect nuclear facilities. Apart from the above design parameters, the structural and functional damage modes resulting from inundation and immersion or current velocity need to be evaluated.

The parameters needed for tsunami probabilistic safety analysis are the same as the design parameters, but the median and variance of the parameter values are required in order to consider its uncertainties.

10.2. ASSESSMENT OF LOAD EFFECTS

Some of the parameters, such as maximum current velocity, inundation depth, maximum negative amplitude or duration, can be estimated using the numerical simulations described in Section 8. For the computation of the inundation depth, detailed information on the plant site is needed. In this section, procedures to evaluate other parameters and their uncertainties are briefly introduced.

It is important to note that the parameters based on experimental data may represent a conservative estimation, and the median and variance of the parameters may be a re-evaluation of the original experimental data.

In 2009, a special issue of *Journal of Disaster Research* [203] compiled 14 papers addressing tsunami effects on buildings and infrastructure. The four main groupings begin with two papers on tsunami force acting on vertical walls. Arikawa [204] experimentally investigates the structural performance of wooden and concrete walls using a large scale laboratory tank in Japan. Also using a similar large scale tsunami flume in the USA, Oshnack et al. [205] studied force reduction by small onshore seawalls in front of a vertical wall. The second grouping focuses on tsunami force on 3-D structures. Arnason et al. [206]

present a basic laboratory study on the hydrodynamics of bore impingement on a vertical column. Fujima et al. [207] examine the two types of formulas for tsunami force evaluation: one is calculated from flow depth alone and the other is based on the Euler number. Lukkunaprasit et al. [208] demonstrate the validity of force computation recommended in a design guideline [209] by the US Federal Emergency Management Agency.

10.2.1. Effects of tsunamis

The effects of tsunamis on structures can be classified into direct and indirect effects. The direct effects of tsunamis on coastal and marine structures can be extensive and often disastrous. Tsunami waves can (i) move entire structures off of their foundations and carry them inland; (ii) damage buildings through impact with vessels carried from offshore and other debris accumulated as the wave advances inland; (iii) undercut foundations and pilings with erosion caused by receding waves; (iv) overturn structures by the suction of receding waves or thrust of advancing waves; and (v) cause large ships to collide with docks during oil or cargo transfer operations, often causing fires. The damage can be quite unexpected [210].

Indirect effects of tsunamis include the resonant oscillations in lakes, basins and harbours, as tsunami periods are often in the range of resonant frequencies of large enclosed or semi-enclosed water bodies. This process is also referred to as sloshing or seiches. Even small tsunamis can trigger resonance and can cause damage to small craft, even in small marinas (see Section 3).

Impact forces can cause a collapse of coastal structures. This process has been observed in excellent detail in photos and videos of the tsunamis of 26 December 2004 and 11 March 2011.

10.2.2. Hydrodynamic force

10.2.2.1. Based on flow velocity

Existing guidelines based on force calculation are referred to as the design flow elevation. For tsunamis, the design flow elevation would be interpreted as the overland flow depth. The associated design velocity is then calculated from $V = 2\sqrt{gd_s}$, that is, twice the long wave celerity at a depth of d_s. This formulation might be conservative because even bores slow down as they advance up a beach on dry land. The US Federal Emergency Management Agency's Coastal Construction Manual (CCM) recommends that this velocity be used to calculate drag forces on piles.

To calculate the hydrodynamic load on a rectangular structure, the CCM recommends two different approaches. When the velocity V calculated as described above is $V < 3.3$ m/s, an equivalent 'dynamic' flow depth is calculated from:

$$d_{dyn} = \frac{1}{2} C_d V^2 / g \tag{34}$$

Synolakis [211] proposed the following relation for hydrodynamic force:

$$F = 2\rho C_d g d_s^2 \tag{35}$$

where C_d is the drag coefficient and depends on the relative ratio of the width of the structure b to the design flow elevation depth d_s at the front of the structure.

For overland flow velocities $V > 3.3$ m/s, the CCM recommends that the force be calculated by:

$$F = \frac{1}{2} \rho C_d V^2 d_s b \tag{36}$$

The CCM describes a methodology for calculating debris impact forces through the calculation of the impact load F_p, given by:

$$F_p = \omega V / gT \tag{37}$$

Here, ω is the weight of the object impacting the structure, V is its velocity, g is the acceleration of gravity and T is the duration of impact. The CCM recommends that, in the absence of any criteria, ω = 1000 lb (453.59 kg) with $V = \sqrt{gd_s}$ be used. Assuming that one is careful with the calculation of the design flow elevation, the formula should produce a conservative estimate [210].

10.2.2.2. Based on flow pressure

Estimation methods of the wave force due to tsunami can be classified into two types according to the location of a physical object. The first type includes offshore structures such as breakwaters and seawalls, and the second type includes onshore structures such as buildings and tanks.

(a) Offshore structures

A tsunami's wave shape in the shallow sea area generally depends on the offshore topography. Without a shoaling beach that is a deep-sea area close to the coast, the tsunami changes the sea level slowly like a quick tidal wave at the coastal zone. In the shoaling beach, the leading edge of the tsunami develops, and its shape changes to a bore or undular bore.

In the former case (i.e. without a shoaling beach), tsunami force acting on a structure such as a breakwater could be estimated as a static hydraulic pressure distribution due to water level change. In the case of a shoaling beach, tsunami wave force acting on the object could be estimated as a total of both kinematic and potential (hydrostatic) energy.

The 1983 Nihonkai-Chubu earthquake caused heavy damage to offshore breakwaters that were under construction at a thermal power plant at Noshiro Port in Niigata, Japan. During this event, an undular bore (or tsunami soliton fission) was observed on the gentle seabed slope off Noshiro [212]. When an undular bore occurs, the new leading wave height develops and breaks. Tanimoto et al. [213] carried out a hydraulic model test and proposed an estimation method for wave pressure distributions after breaking undular bores [214, 215]. Ikeno et al. [216] carried out a hydraulic model test using a variety of tsunami input wave conditions and wave pressure distributions immediately at the beginning of the breaking undular bore.

The pressure intensity P_1 under the still water level and the crest tsunami height η^* can be calculated as follows (see Fig. 66):

$$P_1 = \alpha_B \rho g a_I \tag{38}$$

$$\eta^* = 3.0 a_I \tag{39}$$

$$P_2 = \left(1 - \frac{h^*}{\eta^*}\right) P_1 \tag{40}$$

$$P_u = P_1 \tag{41}$$

$$h^* = min\left(h_C,\ \eta^*\right) \tag{42}$$

$\alpha_B = 2.2$; after breaking [213] (43)

$C_B = 3.0$; immediately at the beginning of breaking [216] (44)

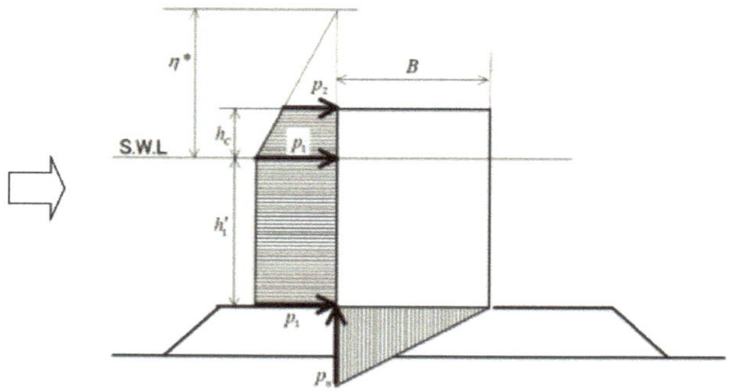

FIG. 66. Wave pressure distributions for an offshore structure. Reproduced courtesy of Japan Society of Civil Engineers [217].

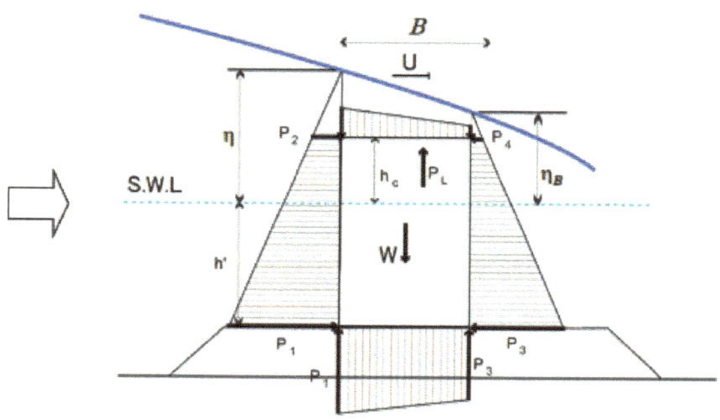

FIG. 67. Tsunami pressure distribution in tsunami overflowings. Reproduced courtesy of Ministry of Agriculture, Forestry and Fisheries, Government of Japan [218].

where ρ is the water density, g is gravity, α_I is the input wave height of the bore, B is the width of the breakwater and α_B is the intensity coefficient. The intensity coefficient may be conservatively taken as Eqs (40) and (41).

Tsunami force acting on a structure in tsunami overflowing could be evaluated by considering the balance of hydrostatic pressure [218] (see Fig. 67), as follows:

Force acting on the front side: P_I

$$P_1 = \rho g(h' + \eta)a_I \quad (45)$$

$$P_2 = \frac{\eta - h_C^*}{h' - \eta} P_1, \; h_C^* = min(\eta, h_C) \tag{46}$$

$$P_I = \frac{1}{2}(P_1 + P_2)(h' + h_C^*) \tag{47}$$

Force acting on the back side: P_B

$$P_3 = \rho g (h' + \eta_B) a_{IB} \tag{48}$$

$$P_4 = P_3 \frac{\eta_B - h_{CB}^*}{h' + \eta_B}, \; h_{CB}^* = min(\eta_B, h_C) \tag{49}$$

$$P_B = \frac{1}{2}(P_1 + P_3)(h' + h_{CB}^*) \tag{50}$$

Uplift pressure: P_U

$$P_U = \frac{1}{2}(P_1 + P_3)B - \frac{1}{2}(P_2 + P_4)B \tag{51}$$

Dynamic lift power (to consider only in the case of overflowing): P_L

$$P_L = C_L \frac{1}{2} \rho U^2 B \tag{52}$$

where C_L is the lift coefficient.

(b) Onshore structures (except tanks)

When a tsunami runs up or draws back, it causes a strong flow similar to that of a river. Structures in the tsunami flow on land receive hydrodynamic force from the flow. Research into the hydrodynamic force on land was active in the 1990s [216, 219, 220]. Two kinds of hydrodynamic force were proposed as calculation formulas: hydrostatic and kinetic type force. It should be noted that both formulas consider dynamic force.

Typical formulas of hydrostatic type force (see Fig. 68) are as follows:

$$p(z)=(\alpha_F \eta_{max}-z)\rho g \tag{53}$$

$$P_H=\int_0^{\alpha_F \eta_{max}} p(z)dz=\frac{1}{2}\rho g \alpha_F^2 \eta_{max}^2 \tag{54}$$

where z is the vertical height from the bottom of the object, $p(z)$ is the pressure dependent on z, P_H is the hydrodynamic force per unit width, α_F is the intensity of the wave pressure, which depends on $F_r=u/\sqrt{g\eta_{max}}$ (Froude number), and u and η_{max} are the horizontal velocity and maximum water depth of a passing wave at the structure. A passing wave means that the wave travels without an object and does not interact with an object.

Asakura et al. [220] proposed the following formulas by compiling experimental data. Variations of α are distributed within $\pm 50\%$.

$$\alpha_F=1.2F_r+1.0(\pm 50\%)\ (0.1\le F_r\le 1.6) \tag{55}$$

$$F_r=\frac{u}{\sqrt{g\eta_{max}}} \tag{56}$$

where F_r is the Froude number.

Sakakiyama [221] rechecked the experimental data of Asakura et al. [220], carried out numerical model testing using 3-D numerical simulation

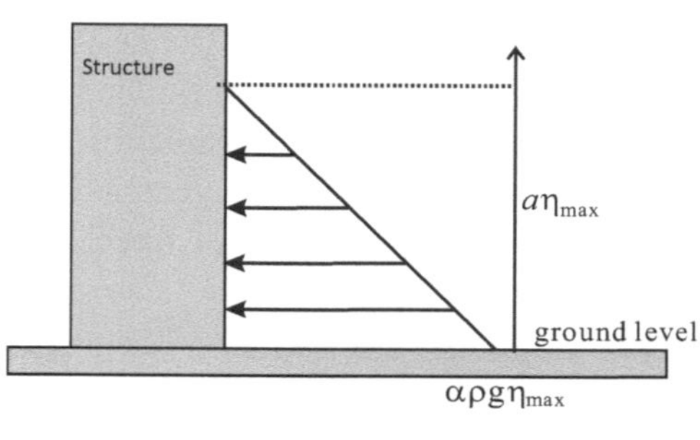

FIG. 68. Wave pressure distribution to an onshore structure.

CADMAS-SURF/3D and proposed new formulas where the variation of α is distributed within –22% $\approx$+12% in the numerical results, as follows:

$$\alpha_{\mathrm{F}} = 1.4F_r + 1.0 \ (-22\% \approx +12\%) \ (0.0 \le F_r \le 2.0) \tag{57}$$

A typical formula of the kinetic type [209, 219, 222] is as follows:

$$P_H = \frac{1}{2} C_D \rho \eta_{max} u^2 \tag{58}$$

where C_D is the drag coefficient (a conservative value is 2.0) and u is the velocity. Equations (54) and (58) have a similar relation through the Froude number (see Eq. (56)).

Impulsive forces are caused by the leading bore edge. The impulsive forces are larger, and the affecting time is instantaneous and shorter, than the tsunami force in the flow in general. It is recommended that the impulsive force P_I be taken as 1.5 times the hydrodynamic force [209].

$$P_I = 1.5P_H \tag{59}$$

(c) Tanks

The Fire and Disaster Management Agency [223] carried out an experimental test and proposed a new formula of wave force. The horizontal wave force F_{tH} can be calculated as follows:

$$F_{tH} = \frac{1}{2}\int_{-\pi}^{\pi} \rho g \left[h_x^{max}(\theta) \right]^2 R cos\theta \ d\theta \tag{60}$$

$$h_x^{max}(\theta) = \alpha \eta_{max} \sum_{m=0}^{3} p_m cos m\theta \tag{61}$$

$$p_0 = 0.680, p_1 = 0.340, p_2 = 0.015, p_3 = -0.035$$

where θ is the azimuth angle around the tank measured from the direction of tsunami incidence.

Below, α is the intensity of the horizontal wave pressure, which depends on F_r (Froude number):

$$\alpha = \begin{cases} 1.8 & F_r \geq 1.3 \\ 2.0F_r - 0.8 & 1.3 \geq F_r \geq 0.9 \\ 1.0 & 0.9 \geq F_r \end{cases} \tag{62}$$

The vertical wave force F_{tV} can be calculated as follows:

$$F_{tV} = 2\int_0^{\pi} \rho g h_V^{max}(\theta) R^2 cos^2\theta \; d\theta \tag{63}$$

$$h_V^{max}(\theta) = \beta\eta_{max} \sum_{m=0}^{3} q_m cosm\theta \tag{64}$$

$q_0 = 0.720$, $q_1 = 0.308$, $q_2 = 0.014$, $q_3 = -0.042$

where β is the intensity of the vertical wave pressure, which depends on F_r (Froude number).

$$\beta = \begin{cases} 1.2 & F_r \geq 1.3 \\ 0.5F_r + 0.55 & 1.3 \geq F_r \geq 0.9 \\ 1.0 & 0.9 \geq F_r \end{cases} \tag{65}$$

10.2.3. Morphological change

Topography change under tsunami action is caused by sediment transport resulting from large flow. The sediment transport can be evaluated by numerical simulation. Studies evaluating topographical change due to tsunamis around the harbours of nuclear power plants have been made by comparing numerical computations with the results of large scale flume experiments around a harbour on a movable bed model [224, 225].

A typical numerical model of tsunami sediment transport was proposed by Takahashi et al. [226], who evaluated it by dividing the water into a bed load layer near the sea-bottom and a suspended layer above, as follows (see Fig. 69):

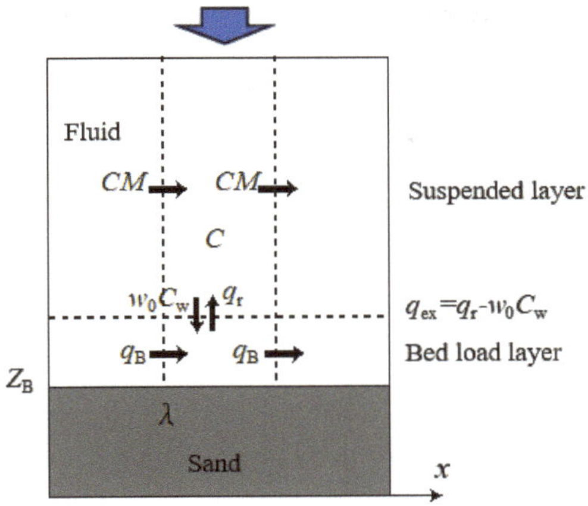

FIG. 69. Suspended and bed load layers [226].

Bed load layer:

$$\frac{\partial Z_B}{\partial t} + \frac{1}{1-\lambda}\left(\frac{\partial q_{Bx}}{\partial x} + \frac{\partial q_{By}}{\partial y} + q_{ex}\right) = 0 \tag{66}$$

Suspended layer:

$$\frac{\partial \bar{C}D}{\partial t} + \frac{\partial CM}{\partial x} + \frac{\partial CN}{\partial y} - q_{ex} = \frac{\partial}{\partial x}\left(K_{hs}\frac{\partial CD}{\partial x}\right) + \frac{\partial}{\partial y}\left(K_{hs}\frac{\partial CD}{\partial y}\right) \tag{67}$$

where Z_B is the height of the bed level above the reference level, λ is porosity, q_{Bx} and q_{By} are the bed load rate in the (x, y) horizontal directions respectively, C is the mean sediment concentration in the suspended layer, and K_{hs} is the horizontal diffusion coefficient. The depth averaged volumetric flux values M and N in the (x, y) horizontal directions, respectively, are calculated using the numerical tsunami model in Section 8.2.1. The tsunami flow drives the concentration of sediment in the suspended layer, whereas the bed load transport depends on the shear force between the fluid and the seabed bottom.

Below, q_{ex} is the sediment exchange rate between the suspended and bed load layers, as follows:

$$q_{ex} = q_r - w_0 C_w \tag{68}$$

where q_r is the pick up rate from the bed load to the suspended layer, w_0 is the settling velocity of the sediment and C_w is the sediment concentration in the boundary layer between suspended and bed load layers.

The formulas for bed load transport q_B and pick up rate q_r are represented as follows:

$$\frac{q_b}{\sqrt{sgd^3}} = a\left(\Psi - \Psi_c\right)^{1.5} \tag{69}$$

$$\frac{q_r}{\sqrt{sgd}} = b\left(\Psi - \Psi_c\right)^2 \tag{70}$$

$$\Psi = \frac{u_*^2}{sgd} = \frac{1}{sgd}\frac{gn^2}{D^{1/3}}\left(M^2 + N^2\right) \tag{71}$$

where s is the density of the sediment in the water, g is gravity, d is the grain size of the sediment, Ψ is Shields' number, which represents dimensionless tractive force, Ψ_c is the critical Shields' number, u_* is the friction velocity, n is Manning's roughness coefficient and a and b are coefficients. Takahashi et al. [226] proposed a, b and C_w as follows:

$$\left(a,\ b,\ C_w\right) = \left(31.0,\ 0.012,\ C\right) \tag{72}$$

Ikeno et al. [227] extended the formulas by dimensional analysis and experimental work as follows [228]:

$$\left(a,\ b,\ C_w\right) = \left(17\left(1 - \frac{\Psi_c}{\Psi}\right)\left(1 - \sqrt{\frac{\Psi_c}{\Psi}}\right),\ 0.15\left(\frac{v^2}{sgd^3}\right)^{0.2}\left(\frac{w_0}{\sqrt{sgd}}\right)^{1.6},\ C_b\right) \tag{73}$$

The bed load transport equation has mainly been applied to sediment transport in river flow [229]. The pick up rate q_{rD} depends on not only Shields' number but also grain size d and settling velocity w_0; and C_b is the concentration at the height of the boundary layer.

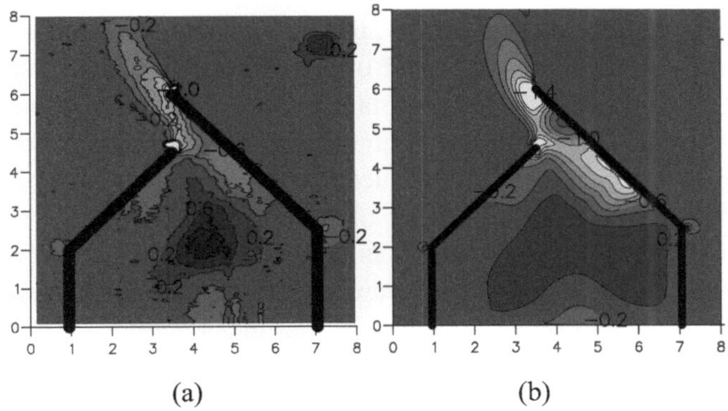

FIG. 70. *Experimental (a) and numerical (b) results of topography change. Reproduced courtesy of Japan Society of Civil Engineers [224].*

$$C_b = \frac{w_0 / a\kappa u_*}{1.0 - exp\left(-w_0 / a\kappa u_*\right)} C \tag{74}$$

where κ is the Kármán constant and a is a constant coefficient (0.2).

A large scale (58 m long, 20 m wide and 1.6 m deep) experiment on tsunami flow around a harbour with a movable bed indicated topographical change [224]. The tsunami caused erosion around the mouth and deposition at the central area of the harbour (see Fig. 70(a)). Figure 70 shows a comparison of experimental and numerical results for topographical change. The numerical model reproduced both erosion near the mouth and sedimentation in the central area of the harbour.

The above-mentioned numerical simulation is a 2-D model that is used extensively for the safety assessment of nuclear power plants in Japan. On the other hand, 3-D numerical simulation has been developing recently. Kihara et al. [225] carried out a 3-D numerical simulation on tsunami induced topography changes near a harbour (the same experiment as shown in Fig. 70). Sediment transport processes on a significant local deposition near the centre of the harbour caused by a tsunami, which was observed in the above-mentioned experimental study, are investigated. This local deposition has not been well predicted by 2-D models.

10.2.4. Collision of drifting debris

Structural damage due to a tsunami is classified into two types: (i) damage caused by the hydrodynamic force of the tsunami itself and (ii) damage caused by drifting debris created by the tsunami. Ships, cars or destroyed objects could become drifting debris. As seen in the 2011 Great East Japan tsunami, drifting debris caused by the tsunami impacted on the coastal area and destroyed both offshore and land structures. The importance of this kind of tsunami effect was realized only recently. Therefore, it is important to develop a tool for predicting the behaviour of the drifting debris produced by tsunamis at nuclear power plants.

Numerical models of the spread of timber in tsunamis were developed by Goto [230]. The proposed model was applied to Miyako Bay on the North Sanriku coast of Japan. Goto [230] assumed that the motion of timber consists of two parts: mean motion and diffusion. The mean motion is governed and determined by the mean current induced by tsunamis when the drag and virtual mass effects of timber are combined. The mean motion of timber is governed by the following equations:

$$\rho_t V \frac{dU_t}{dt} = \rho V \frac{dU}{dt} + \rho(C_M - 1)V\left(\frac{dU}{dt} - \frac{dU_t}{dt}\right) - \frac{1}{2}\rho C_D A(U_t - U)|U_t - U| \tag{75}$$

where U_t and U are the speed vectors of timber and water particles, respectively, and ρ_t and ρ express the density of timber and water. The virtual mass coefficient C_M and the drag coefficient C_D are determined by hydraulic experiments, as follows:

$$C_M = 1.78 \tag{76}$$

$$C_D = \left(0.91 + 32.5R_e^{-1/2}\right)^2 + 0.1F_r^{1/4} \tag{77}$$

where R_e and F_r indicate the Reynolds and Froude numbers of the timber spread.

The diffusion is determined by a consideration that the spread of timber around its centre of gravity follows a random process, the variance of the probability density of which is equivalent to the magnitude of diffusion obtained by experiments. The position of timber X is expressed in following equation:

$$X = \int_0^t U_t dt + \sum_{k=0}^{n\Delta t} \sqrt{24k\Delta t}\left(\xi^t - 0.5\right) \tag{78}$$

where ξ^t is given as a random function uniformly distributed in the interval -1 to 1. Here, k denotes the diffusion coefficient determined by the hydraulic experiment as in the following equation:

$$k = 0.0032U^*h \tag{79}$$

where U^* is the friction velocity of the water flow and h is the water depth.

The numerical model for the drift behaviour of containers and their collision force was proposed by Kumagai et al. [231]. An outline of the model constructed is shown in Fig. 71. First, a horizontal 2-D numerical simulation of a tsunami is carried out to obtain the water depth and flow velocity distribution in the object area, and the drifted container is evaluated on the basis of the characteristics of the tsunami. The behaviour of the container is then calculated by solving the

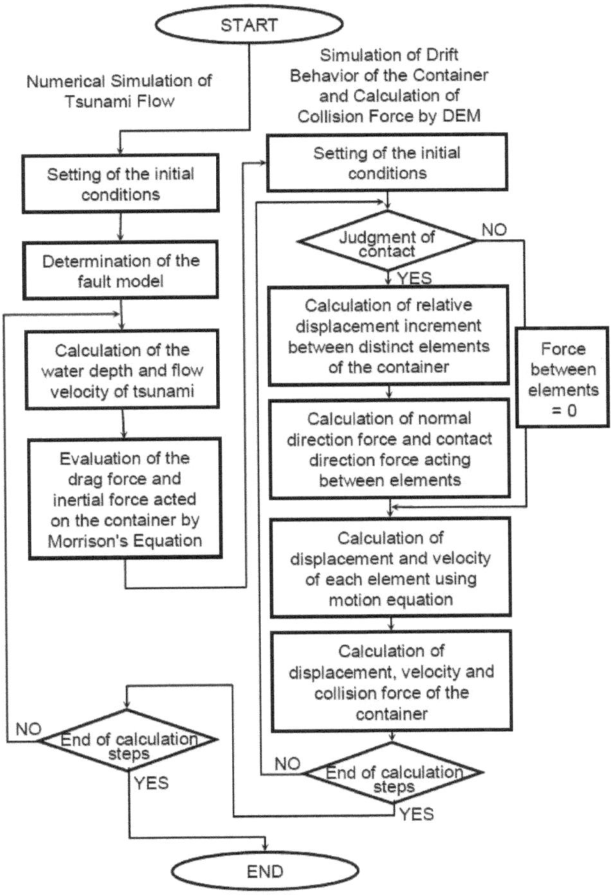

FIG. 71. Outline flowchart for the collision of drifting debris model by Kumagai et al. DEM denotes the distinct element method. Reproduced courtesy of Techno-Ocean Network [231].

motion equations for the container. To estimate a magnitude of collision force between the container drifted by the tsunami and a coastal structure, such as a tide wall, the container is expressed by several elements using the distinct element method. The effect of rotation and other behaviour of the container using the distinct element method is also considerable. As a result, the collision force between debris and structures can be calculated.

Fujii et al. [232] carried out numerical simulations for experimental tests of drifting ships by connecting several elements (extended distinct element method). As a result, it was confirmed that the numerical model can well reproduce the behaviour of drifting bodies due to tsunamis, including complex behaviour such as rotation in a harbour, running on land and running over a breakwater.

When oil storage tanks drift from their initial positions as a result of a tsunami, they may cause severe problems, such as the destruction of structures nearby and leakage of stored oil. Inagaki et al. [233] carried out experiments on the drifting of tanks caused by a tsunami. These revealed that the tsunami wave force and friction forces determine the slipping and drifting of the tanks. Inagaki et al. [233] constructed a simulation model using the momentum equation for the prediction of the tank movement.

10.2.5. Deterministic and probabilistic estimates

Some of the parameters can be estimated from empirical relations with water height or flow depths. All of the above parameters can be estimated from the numerical modelling described in Section 8, with additional information on plant grade, the layout of structures, etc. For this computation, the tsunami's source parameters must be specified.

Section 9 describes the probabilistic approach (i.e. PTHA) used to obtain a hazard curve, a curve for probability or an annual exceedance rate for different tsunami amplitudes. The design tsunami amplitude or runup can be estimated from this curve for a specific design probability. The design probability for the safety of nuclear facilities can be specified by individual Member States. It is ideal to conduct the PTHA for other parameters, but it is often difficult. If the deaggregation of the PTHA shows that a particular source (e.g. an earthquake source with a certain magnitude) is the major contributor, the other parameters can be estimated by deterministic analysis of the dominant event. However, the above scenario event may not provide the design parameters and may instead provide several sets of parameters for further deterministic studies.

11. MANAGEMENT SYSTEM

11.1. APPLICATION OF THE MANAGEMENT SYSTEM

A management system applicable to all organizations involved in the evaluation of tsunami and seiche hazards should be established and implemented before the hazard evaluation programme. Guidelines on conducting the application of the management system are contained in Section 11 of SSG-18 [1].

The management system should cover all processes and activities of this programme, in particular those relating to data collection, the specification and parameterization of sources, propagation analysis and hazard assessment that lie within the scope of the present report. It should also cover those processes and activities corresponding to the re-evaluation phase of the programme.

Similarly, a peer review of how the investigation and analytical methodology were implemented needs to be performed. In particular, the peer review should assess the evaluation of tsunami and seiche hazards against the recommendations of the present report and current international good practice used for these evaluations.

The peer review is conducted by competent experts and needs to be properly documented, as explained below.

11.2. DOCUMENTATION AND RECORDS

Documentation of actions taken, including the responsible party and justification, should be performed promptly (see Section 12). Such documentation is essential for reconstructing the data, methods and procedures used for the evaluation of tsunami and seiche hazards. The documentation serves many purposes, such as providing lessons learned for future impacts, as well as the timing and substance of communication to governing bodies, the public and other stakeholders.

Therefore, an important component of the management system is the definition of the documentation and records to be developed during the implementation of the different steps of the programme, the execution of the evaluation of tsunami and seiche hazards, and the final report to be produced as a result. Detailed documentation is retained for review and future application. The documentation should identify all sources of information used in the hazard analysis, including information on where to find important citations that may be difficult to obtain. Unpublished data that are used in the analysis should be included in the documentation in an appropriately accessible and usable form. Documentation should be detailed enough to allow for independent review.

11.3. SPECIFIC QUALITY ASSURANCE AREAS

Preparation of evaluation plan:

— The evaluation plan should be prepared with consideration of the objectives and content of the evaluation of tsunami and seiche hazards as well as the implementation period.

Data collection:

— The quality and resolution of the database that is used have to be commensurate with the intended use in tsunami and seiche propagation and runup analyses.
— The sources of the data (prehistorical data, historical data, field survey results and instrumental data) should be clearly defined and well established.
— In consideration of the precision necessary for the evaluation, appropriately precise data should be collected. The adequacy of the data should be established.

Specification and parameterization of tsunami and seiche sources:

— There should be a range of source models in the coverage of analytical methodologies.
— The ideas of the source models should be demonstrated, and the source models should be specified based on the objectives of the evaluation.

Tsunami and seiche propagation analyses:

— The analytical methodologies and analytical cases should be selected on the basis of the objectives and content of the evaluation and implementation period. When the influences of uncertain factors are assumed, sensitivity analyses should be planned as needed.
— Rational analytical methodologies should be selected with consideration of the implementation period, required time and expense of analyses.
— In the selection of analytical methodologies, it should be confirmed that the selected items of each analytical methodology satisfy the requirements of the objectives and content of the evaluation.

Verification and validation:

- The results of the analysis should be compared with the obtained data from the applied site, experimental results and theoretical values, and the range of applied selected analytical methodologies should be confirmed. The conditions of verification should be clarified.
- When the operational checks for analytical functions are difficult to undertake using complex analytical models, the operational checks should be confirmed using simple analytical models.
- The trends of the analytical results should be compared using a similar phenomenon, and the verification of analytical results should be confirmed.
- Regarding analytical parameters with uncertainties, sensitivity analyses that encompass the distributions of parameters should be carried out, and the effects for analytical results should be confirmed.
- The adequacy of analytical results should be confirmed using a checklist, and the result of these confirmations should be recorded.

Verification and validation documents for the numerical model covering the above requirements for benchmark problems may be submitted separately.

Tsunami and seiche hazard assessments of coastal impacts:

- Periodic meetings should be held by the team implementing the programme of evaluation. The status of progress, the presence or absence of problems and the policies to address them should also be discussed thoroughly by the team.
- Regarding those items of evaluation results where uncertainties are assumed, the sensitivities should be analysed, and the ranges of uncertainties should be stated.
- If earlier studies on the evaluation of tsunami and seiche hazards are available, comparisons should be made to demonstrate how the use of different approaches or different data affects the conclusions. The comparisons should be documented in a way that allows for their review.

Peer review:

- The level and type of peer review can vary depending on the safety category of the structure located at the site. The peer review should address all parts of the hazard evaluation, including the process for the evaluation of tsunami and seiche hazards, all technical elements (e.g. tsunami and seiche source characterizations, propagation models, etc.), the method of evaluation of tsunami and seiche hazards, and quantification and documentation.

— The peer review should be conducted by selected experts in the fields of seismology, geology, oceanography, hydrology, volcanology, earthquake and geotechnical engineering, and civil and coastal engineering.
— The results of the peer review should be properly documented.

12. DOCUMENTATION

12.1. GENERAL

To make the hazard evaluation traceable and transparent to users, peer reviewers, the licensee and the regulatory body, the documentation of evaluation of tsunami and seiche hazards should provide the following:

- A description of all elements of the process for the evaluation of tsunami and seiche hazards;
- The identities of the study participants and their roles;
- Background material that comprises the analyses, including raw and processed data, computer software and the input and output files, reference documents, the results of intermediate calculations and sensitivity studies.

12.2. RECOMMENDED DOCUMENT STRUCTURE

A recommended document structure may be as follows:

(a) Introduction.
(b) Identification of the study participants and their roles.
(c) Background information for the assessment:
   (i) Assessment area;
   (ii) Nuclear power plant information;
   (iii) Quality control procedures for assessment.
(d) Assessment flow.
(e) Data collection:
   (i) Catalogue of past tsunamis (prehistorical data, historical data, field survey results and instrumental data) and detailed reconstruction of any past events;
   (ii) Description of local and regional geology (with recognition of evidence of relevant capable faults and landslide and volcanic activity), seismological data and seismogenic sources;
   (iii) Topography and bathymetry data.
(f) Specification and parameterization of tsunami and seiche sources:
   (i) Target domain;
   (ii) Selection of source models from the possible source mechanisms;
   (iii) Specification of parameters for source models;
   (iv) Uncertainties.

(g) Tsunami and seiche propagation analyses:
   (i) Selection of numerical analysis;
   (ii) Specification of parameters for numerical analysis;
   (iii) Uncertainties.
(h) Hazard assessment of coastal impacts:
   (i) Selection of hazard assessment methodologies;
   (ii) Simulation results;
   (iii) Analysis, comparison, presentation and discussion of simulation results;
   (iv) Determination of critical design parameters from simulation results;
   (v) Uncertainties.
(i) Conclusion:
   (i) Conclusion for the objective;
   (ii) Findings.
(j) Appendix:
   (i) Raw and processed data;
   (ii) Computer software, and the input and output files;
   (iii) Reference documents;
   (iv) Results of intermediate calculations;
   (v) Sensitivity studies;
   (vi) Verification and validation studies.

REFERENCES

[1] INTERNATIONAL ATOMIC ENERGY AGENCY, WORLD METEOROLOGICAL ORGANIZATION, Meteorological and Hydrological Hazards in Site Evaluation for Nuclear Installations, IAEA Safety Standards Series No. SSG-18, IAEA, Vienna (2011).

[2] INTERNATIONAL ATOMIC ENERGY AGENCY, The Fukushima Daiichi Accident — Report by the Director General, Non-serial Publications, IAEA, Vienna (2015),
https://www-pub.iaea.org/mtcd/publications/pdf/pub1710-reportbythedg-web.pdf

[3] IIDA, K., COX, D.C., PARARAS-CARAYANNIS, G., Preliminary Catalog of Tsunamis Occurring in the Pacific Ocean, HIG-67-10, University of Hawaii, Honolulu (1967),
http://www.soest.hawaii.edu/Library/Tsunami%20Reports/Iida_et_al.pdf

[4] HATORI, T., Classification of tsunami magnitude scale, Bull. Earthq. Res. Inst. Univ. Tokyo **61** (1986) 503–515 (in Japanese with English abstract),
https://cir.nii.ac.jp/crid/1390290701029981184

[5] SOLOVIEV, S.L., GO, C.N., A Catalogue of Tsunamis on the Western Shore of the Pacific Ocean, Nauka Publishing House, Moscow (1974) (in Russian) [Canadian translation: Can. J. Fish. Aquat. Sci. **5077** (1984)].

[6] PORT AND AIRPORT RESEARCH INSTITUTE, Preliminary Report of Tsunami and Earthquake Damage of Port, Coast and Airport by 2011 Great East Japan Earthquake, Tech. Note No. 1231, Port and Airport Research Institute, Yokosuka (2011).

[7] HILDEBRAND, A.R., et al., Chicxulub crater: A possible Cretaceous/Tertiary boundary impact crater on the Yucatán Peninsula, Mexico, Geology **19** (1991) 867–871, https://doi.org/10.1130/0091-7613(1991)019%3C0867:CCAPCT%3E2.3.CO;2

[8] MATSUI, T., IMAMURA, F., TAJIKA, E., NAKANO, Y., FUJISAWA, Y., "Generation and propagation of a tsunami from the Cretaceous/Tertiary impact event", Catastrophic Events and Mass Extinctions: Impact and Beyond, Geological Society of America, Boulder, CO (2002) 69–72.

[9] KÂNOGLU, U., et al., Focusing of long waves with finite crest over constant depth, Proc. R. Soc. London, Ser. A Math Phys. Sci. **469** (2013) 20130015, https://doi.org/10.1098/rspa.2013.0015

[10] TADEPALLI, S., SYNOLAKIS, C.E., The run-up of *N*-waves on sloping beaches, Proc. R. Soc. London, Ser. A Math Phys. Sci. **445** (1994) 99–112, https://doi.org/10.1098/rspa.1994.0050

[11] TADEPALLI, S., SYNOLAKIS, C.E., Model for the leading waves of tsunamis, Phys. Rev. Lett. **77** (1996) 2141–2144, https://doi.org/10.1103/PhysRevLett.77.2141

[12] RABINOVICH, A.B., "Seiches and harbor oscillations", Handbook of Coastal Engineering (KIM, Y.C., Ed.), World Scientific Publications, Singapore (2009) Ch. 9.

[13] THE OVERSEAS COASTAL AREA DEVELOPMENT INSTITUTE OF JAPAN, Technical Standards and Commentaries for Port and Harbour Facilities in Japan, OCDI, Tokyo (2002).

[14] NIEDDA, M., GREPPI, M., Tidal, seiche and wind dynamics in a small lagoon in the Mediterranean Sea, Estuar. Coast. Shelf Sci. **74** (2007) 21–30, https://doi.org/10.1016/j.ecss.2007.03.022

[15] HONDA, K., TERADA, T., YOSHIDA, Y., ISITANI, D., An investigation on the secondary undulations of oceanic tides, J. Coll. Sci., Imp. Univ. Tokyo (1908) 108.

[16] BARBEROPOULOU, A., QAMAR, A., PRATT, T.L., CREAGER, K.C., STEELE, P., Local amplification of seismic waves from the Denali Earthquake and damaging seiches in Lake Union, Seattle, Washington, Geophys. Res. Lett. **31** (2004) L03607, https://doi.org/10.1029/2003GL018569

[17] BARBEROPOULOU, A., QAMAR, A., PRATT, T.L., STEELE, P., Long-period effects of the Denali earthquake on water bodies in the Puget lowland: observations and modeling, Bull. Seismol. Soc. Am. **96** 2 (2006) 519–535, https://doi.org/10.1785/0120050090

[18] BARBEROPOULOU, A., A seiche hazard study for Lake Union, Seattle, Washington, Bull. Seismol. Soc. Am. **98** 4 (2008) 1837–1848, https://doi.org/10.1785/0120070153
[19] BRUUN, P., Port Engineering, Gulf Publishing, Houston and London (1981) 94–108.
[20] YALCINER, A.C., PELINOVSKY, E., A short cut numerical method for determination of resonance periods of free oscillations in irregular shaped basins, Ocean Eng. **34** 5–6 (2007) 747–757, https://doi.org/10.1016/j.oceaneng.2006.05.016
[21] RAICHLEN, F., "Harbor resonance, interaction of structures and waves", Estuary and Coastline Hydrodynamics (IPPEN, A.T., Ed.), McGraw-Hill, New York (1966) 281–315.
[22] RAICHLEN, F., LEE, J., "Oscillation of bays, harbors, and lakes", Handbook of Coastal and Ocean Engineering (HERBICH, J., Ed.), Gulf Publishing, Houston (1991).
[23] LEE, J., Wave-induced oscillations in harbours of arbitrary geometry, J. Fluid Mech. **45** Part 2 (1971) 375–394, https://doi.org/10.1017/S0022112071000090
[24] YAO, L.S., A resonant wave theory, J. Fluid Mech. **395** (1999) 237–251, https://doi.org/10.1017/S0022112099005832
[25] LEE, J., RAICHLEN F., Oscillations in harbors with connected basins, J. Waterw. Harbors Coast. Eng. Div. **98** (1972) 311-332, https://doi.org/10.1061/AWHCAR.0000153
[26] JAPAN NUCLEAR ENERGY SAFETY ORGANIZATION, Annual Report 2010, JNES, Tokyo (2011).
[27] IMAMURA, F., et al., Field survey of the 1994 Mindoro Island, Philippines tsunami, Pure Appl. Geophys. **144** 3–4 (1995) 875–890, https://doi.org/10.1007/BF00874399
[28] UNITED STATES GEOLOGICAL SURVEY, M 9.1 — 2004 Sumatra — Andaman Islands Earthquake, https://earthquake.usgs.gov/earthquakes/eventpage/official20041226005853450_30/executive
[29] SHETH, A., SANYAL, S., JAISWAL, A., GANDHI, P., Effects of the December 2004 Indian Ocean tsunami on the Indian mainland, Earthq. Spectra **22** S3 (2006) S435–S473, https://doi.org/10.1193/1.2208562
[30] FUJII, Y., SATAKE, K., Tsunami source of the 2004 Sumatra–Andaman earthquake inferred from tide gauge and satellite data, Bull. Seismol. Soc. Am., Special Issue, (2007) 192–207, https://doi.org/10.1785/0120050613
[31] GRILLI, S.T., et al., Source constraints and model simulation of the Dec. 26, 2004, Indian Ocean tsunami, J. Waterw. Port Coast. Ocean Eng. **133** 6 (2007) 414–428, https://doi.org/10.1061/(ASCE)0733-950X(2007)133:6(414)

[32] HERBERT, H., SLADEN, A., SCHINDELE, F., Numerical modelling of the great 2004 Indian Ocean tsunami: Focus on the Mascarene Islands, Bull. Seismol. Soc. Am. **97** 1A (2007) S208–S222,
https://doi.org/10.1785/0120050611

[33] PIATANESI, A., LORITTO, S., Rupture process of the 2004 Sumatra–Andaman earthquake from tsunami waveform inversion, Bull. Seismol. Soc. Am. **97** (2007) 223–231,
https://doi.org/10.1785/0120050627

[34] UNITED STATES GEOLOGICAL SURVEY, Significant Earthquake — 2004,
https://earthquake.usgs.gov/earthquakes/eventpage/official20041226005853450_30/impact

[35] NAGARAJAN, B., et al., The great tsunami of 26 December 2004: A description based on tide gauge data from the Indian subcontinent and surrounding areas, Earth Planets Space **58** 2 (2006) 211–215,
https://doi.org/10.1186/BF03353380

[36] NATIONAL INSTITUTE OF OCEANOGRAPHY, Tide-gauge Observations of the 2004 Indian Ocean Tsunami, NIO, India (2012),
https://publication-data.nio.res.in/s/pWebS3xJfwaZHN6

[37] KRISHNAMOORTY, S., "Effect of tsunami on Madras atomic power station (MAPS), December 26, 2004", paper presented at IAEA Int. Workshop on External Flooding Hazards at Nuclear Power Plant Sites, Kalpakkam, India, 2005.

[38] GOVERNMENT OF INDIA, National Report to the Convention on Nuclear Safety, Fourth Review Meeting of Contracting Parties, April 2008,
https://www.aerb.gov.in/images/PDF/IndianNationalReportCNS.pdf

[39] NUCLEAR POWER CORPORATION OF INDIA, Safety Evaluation of Indian Nuclear Power Plants PHWRs at Madras Atomic Power Station, NPCIL, Mumbai (2011).

[40] SINGH, R.K., KUSHWAHA, H.S., "Tsunami hazard evaluation of Indian nuclear coastal sites for Sumatra 2004 event", paper presented at IAEA Int. Workshop on External Flooding Hazards at Nuclear Power Plant Sites in Commemoration of Five Years of Indian Ocean Tsunami, Kalpakkam, India, 2010.

[41] PATHAN, S.K., BHANDERI, R.J., VINOTH KUMAR, J.A., SASIDHAR, P., "Mapping flood inundation patterns in and around Kalpakkam nuclear power plant site using geo-informatics techniques", paper presented at IAEA Int. Workshop on External Flooding Hazards at Nuclear Power Plant Sites, Kalpakkam, India, 2005.

[42] BHARATIYA NABHIKIYA VIDYUT NIGAM, Annual Report for 2005, BHAVINI, Chennai (2005).

[43] KUMAR, P., "Quality assurance in BHAVINI", Proc. Natl. Semin. on Non-Destructive Evaluation, Indian Society for Non-Destructive Testing, Hyderabad Chapter, Hyderabad, India, 2006,
www.ndt.net/article/nde-india2006/files/PL-02.pdf

[44] KUMAR, P., "Preparedness at PFBR Kalpakkam to meet the challenges due to natural events", Proc. 2nd Joint GIF–IAEA/INPRO Workshop on Safety Aspects of Sodium-Cooled Fast Reactors, IAEA, Vienna, 2011,
https://inis.iaea.org/collection/NCLCollectionStore/_Public/49/059/49059940.pdf

[45] KUMAR, P., "Presentation and narration on December 26, 2004 tsunami", paper presented at IAEA Int. Workshop on External Flooding Hazards at Nuclear Power Plant Sites, Kalpakkam, India, 2005.

[46] KUMAR, P., "Experiences of 26th Dec 2004 tsunami at Kalpakkam", paper presented at IAEA Int. Workshop on External Flooding Hazards at Nuclear Power Plant in Commemoration of Five Years of Indian Ocean Tsunami Event, Kalpakkam, India, 2010.

[47] Atomic Energy Regulatory Board, "Seismic Studies and Design Basis Ground Motion for Nuclear Power Plant Sites", AERB/SG/S-11, AERB (1990), https://aerb.gov.in/storage/uploads/documents/regdocWndoQ.PDF

[48] DEPARTMENT OF ATOMIC ENERGY, GOVERNMENT OF INDIA, press release, https://archivepmo.nic.in/drmanmohansingh/press-details.php?nodeid=157

[49] ATOMIC ENERGY REGULATORY BOARD, Annual Report for 2004, AERB, Mumbai (2004), https://www.aerb.gov.in/images/PDF/report/annrpt2k4-min.pdf

[50] GOVERNMENT OF INDIA, National Report (Actions taken for Indian NPPs Subsequent to Fukushima Nuclear Accident) to the Second Extraordinary Meeting of Contracting Parties to the Convention on Nuclear Safety, August 2012, https://www.aerb.gov.in/images/PDF/CNS2012.pdf

[51] ATOMIC ENERGY REGULATORY BOARD, Report of AERB Committee to Review Safety of Indian Nuclear Power Plants Against External Events of Natural Origin, AERB, Mumbai (2011), https://www.aerb.gov.in/storage/images/PDF/09-November-2011.pdf

[52] SATAKE, K., FUJII, Y., HARADA, T., NAMEGAYA, Y., Time and slip distribution of coseismic slip of the 2011 Tohoku earthquake as inferred from tsunami waveform data, Bull. Seismol. Soc. Am. **103** 2B (2013) 1473–1492, https://doi.org/10.1785/0120120122

[53] OZAWA, S., et al., Coseismic and postseismic slip of the 2011 magnitude-9 Tohoku-Oki earthquake, Nature **475** (2011) 373–376, https://doi.org/10.1038/nature10227

[54] IDE, S., BALTAY, A., BEROZA, G.C., Shallow dynamic overshoot and energetic deep rupture in the 2011 Mw 9.0 Tohoku-oki earthquake, Science **332** 6036 (2011) 1426–1429, https://doi.org/10.1126/science.1207020

[55] SATO, M., et al., Displacement above the hypocenter of the 2011 Tohoku-oki earthquake, Science **332** 6036 (2011) 1395, https://doi.org/10.1126/science.1207401

[56] MORI, N., TAKAHASHI, T., YASUDA, T., YANAGISAWA, H., Survey of 2011 Tohoku earthquake tsunami inundation and run-up, Geophys. Res. Lett. **38** 7 (2011) L00G14, https://doi.org/10.1029/2011GL049210

[57] TANG, L., et al., Direct energy estimation of the 2011 Japan tsunami using deep-ocean pressure measurements, J. Geophys. Res. **117** C8 (2012) C08008, https://doi.org/10.1029/2011JC007635

[58] FUJII, Y., SATAKE, K., SAKAI, S., SHINOHARA, M., KANAZAWA, T., Tsunami source of the 2011 off the Pacific coast of Tohoku earthquake, Earth Planets Space **63** 7 (2011) 815–820, https://doi.org/10.5047/eps.2011.06.010

[59] NUCLEAR EMERGENCY RESPONSE HEADQUARTERS, "The accident at TEPCO's Fukushima nuclear power station", Report of the Japanese Government to the IAEA Ministerial Conference on Nuclear Safety, Vienna, 2011, Government of Japan, Tokyo (2011), https://japan.kantei.go.jp/kan/topics/201106/iaea_houkokusho_e.html

[60] NUCLEAR AND INDUSTRIAL SAFETY AGENCY, "The results of the investigation and evaluation regarding tsunami generated by Tohoku-Chihou-Taiheiyo-Oki earthquake", Advisory Committee for Earthquake and Tsunami, 2nd Meeting, NISA, Tokyo, 5 October 2011, (in Japanese), https://warp.ndl.go.jp/info:ndljp/pid/3537352/www.nsr.go.jp/archive/nisa/shingikai/800/26/002/231005.html

[61] NUCLEAR AND INDUSTRIAL SAFETY AGENCY, Technical Knowledge of the Accident at Fukushima Daiichi Nuclear Power Station of Tokyo Electric Power Company (2012), https://warp.da.ndl.go.jp/info:ndljp/pid/3532877/www.nisa.meti.go.jp/english/press/2012/06/en20120615-1.html

[62] NUCLEAR EMERGENCY RESPONSE HEADQUARTERS, The Accident at TEPCO's Fukushima Nuclear Power Station (Second Report), Additional report of the Japanese Government to the IAEA, Government of Japan, Tokyo (2011), https://www.iaea.org/report-japanese-government-iaea-ministerial-conference-nuclear-safety-accident-tepcos-fukushima-nuclear-power-stations

[63] TOKYO ELECTRIC POWER COMPANY, Fukushima Nuclear Accident Analysis Report, Interim Report, TEPCO, Tokyo (2011), http://www.tepco.co.jp/en/press/corp-com/release/11120205-e.html; Final Report, TEPCO, Tokyo (2012), http://www.tepco.co.jp/en/press/corp-com/release/2012/1205638_1870.html

[64] TOKYO ELECTRIC POWER COMPANY, Effects of the Earthquake and Tsunami on the Fukushima Daiichi and Daini Nuclear Power Stations, TEPCO, Tokyo (2011), http://www.tepco.co.jp/en/nu/fukushima-np/images/handouts_110525_01-e.pdf

[65] TOHOKU ELECTRIC POWER COMPANY, Tsunami Evaluation and Measures at Onagawa Nuclear Power Station, Nuclear Safety Commission, Tokyo (2011), https://warp.da.ndl.go.jp/info:ndljp/pid/3533051/www.nsc.go.jp/senmon/shidai/jishin/jishin4/siryo4-2.pdf

[66] INTERNATIONAL ATOMIC ENERGY AGENCY, IAEA Mission to Onagawa Nuclear Power Station to Examine the Performance of Systems, Structures and Components Following the Great East Japanese Earthquake and Tsunami, IAEA, Vienna (2012)
https://inis.iaea.org/search/search.aspx?orig_q=RN:44050829

[67] INTERNATIONAL ATOMIC ENERGY AGENCY, IAEA Report on Protection Against Extreme Earthquakes and Tsunamis in the Light of the Accident at the Fukushima Daiichi Nuclear Power Plant, Action Plan on Nuclear Safety Series, IAEA, Vienna (2012)
https://www-pub.iaea.org/MTCD/Publications/PDF/IEM-3_web.pdf

[68] TOLKOVA E., Tide-tsunami interaction in Columbia River, as implied by historical data and numerical simulations, Pure Appl. Geophys. **170** 6–8 (2013) 1115–1126,
https://doi.org/10.1007/s00024-012-0518-0

[69] INTERNATIONAL ATOMIC ENERGY AGENCY, Technical Approach to Probabilistic Safety Assessment for Multiple Reactor Units, Safety Reports Series No. 96, IAEA, Vienna (2019).

[70] INTERNATIONAL ATOMIC ENERGY AGENCY, The Contribution of Palaeoseismology to Seismic Hazard Assessment in Site Evaluation for Nuclear Installations, IAEA-TECDOC-1767, IAEA, Vienna (2015).

[71] MILLER, D.J., Giant waves in Lituya Bay, Alaska, U.S. Geol. Surv., Prof. Pap. **354-C** (1960) 51–86,
https://pubs.usgs.gov/pp/0354c/report.pdf

[72] FRITZ, H.M., MOHAMMED, F., YOO, J., Lituya Bay landslide impact generated mega-tsunami 50th anniversary, Pure Appl. Geophys. **166** 1–2 (2009) 153–175,
https://doi.org/10.1007/s00024-008-0435-4

[73] PLAFKER, G., et al., Effects of the earthquake of March 27, 1964, on various communities, U.S. Geol. Surv., Prof. Pap. **542-G** (1969) 1–50,
https://doi.org/10.3133/pp542G

[74] SULEIMANI, E., NICOLSKY, D.J., HAEUSSLER, P.J., HANSEN, R., Combined effects of tectonics and landslide-generated tsunami runup at Seaward, Alaska during the Mw 9.2 1964 earthquake, Pure Appl. Geophys. **168** 6 (2011) 1053–1074,
https://doi.org/10.1007/s00024-010-0228-4

[75] SYNOLAKIS, C.E., et al., The slump origin of the 1998 Papua New Guinea tsunami, Proc. R. Soc. London, Ser. A Math. Phys. Sci. **458** (2002) 763–789,
https://doi.org/10.1098/rspa.2001.0915

[76] SATAKE, K., TANIOKA, Y., The July 1998 Papua New Guinea earthquake: Mechanism and quantification of unusual tsunami generation, Pure Appl. Geophys. **160** 10–11 (2003) 2087–2118,
https://doi.org/10.1007/s00024-003-2421-1

[77] SATAKE, K., Volcanic origin of the 1741 Oshima-Oshima tsunami in the Japan Sea, Earth Planets Space **59** 5 (2007) 381–390,
https://doi.org/10.1186/BF03352698

[78] KAWAMATA, K., et al., “Model of tsunami generation by collapse of volcanic eruption: The 1741 Oshima-Oshima tsunami”, Tsunamis: Case Studies and Recent Developments, (SATAKE, K., Ed.), Springer, Dordrecht (2005) 79–96.

[79] KANAMORI, H., Mechanism of tsunami earthquakes, Phys. Earth Planet. Inter. **6** (1972) 346–359,
https://doi.org/10.1016/0031-9201(72)90058-1

[80] SATAKE, K., TANIOKA, Y., Sources of tsunami and tsunamigenic earthquakes in subduction zones, Pure Appl. Geophys. **154** 3–4 (1999) 467–483,
https://doi.org/10.1007/s000240050240

[81] MUROTANI, S., MIYAKE, H., KOKOETSU, K., Scaling of characterized slip models for plate-boundary earthquakes, Earth Planets Space **60** (2008) 987–991,
https://doi.org/10.1186/BF03352855

[82] SATAKE, K., et al., Tsunami source of the 2010 Mentawai, Indonesia earthquake inferred from tsunami field survey and waveform modelling, Pure Appl. Geophys. **170** 9–10 (2013) 1567–1582,
https://doi.org/10.1007/s00024-012-0536-y

[83] BILEK, S.L., LAY, T., Rigidity variations with depth along interplate megathrust faults in subduction zones, Nature **400** (1999) 443–446,
https://doi.org/10.1038/22739

[84] STEKETEE, J.A., On Volterra’s dislocations in a semi-infinite elastic medium, Can. J. Phys. **36** (1958) 192–205,
https://doi.org/10.1139/p58-024

[85] OKADA, Y., Surface deformation due to shear and tensile faults in a half-space, Bull. Seismol. Soc. Am. **75** 4 (1985) 1135–1154,
https://doi.org/10.1785/BSSA0750041135

[86] MANSINHA, L., SMYLIE, D.E., The displacement fields of inclined faults, Bull. Seismol. Soc. Am. **61** 5 (1971) 1433–1440,
https://doi.org/10.1785/BSSA0610051433

[87] KAJIURA, K., The leading wave of a tsunami, Bull. Earthq. Res. Inst. Univ. Tokyo **41** (1963) 535–571,
https://doi.org/10.15083/0000033711

[88] AIDA, I., Numerical experiments for the tsunami propagation — The 1964 Niigata tsunami and the 1968 Tokachi-oki tsunami, Bull. Earthq. Res. Inst. Univ. Tokyo **47** (1969) 673–700,
https://doi.org/10.15083/0000033361

[89] TANIOKA, Y., SATAKE, K., Tsunami generation by horizontal displacement of ocean bottom, Geophys. Res. Lett. **23** (1996) 861–864,
https://doi.org/10.1029/96GL00736

[90] AIDA, I., Reliability of a tsunami source model derived from fault parameters, J. Phys. Earth **26** (1978) 57–73,
https://doi.org/10.4294/jpe1952.26.57

[91] JAPAN SOCIETY OF CIVIL ENGINEERS, Tsunami Assessment Method for Nuclear Power Plants in Japan 2016, JSCE, Tokyo (2016), https://committees.jsce.or.jp/ceofnp/node/140

[92] MCCAFFREY, R., Global frequency of magnitude 9 earthquakes, Geology **36** 3 (2008) 263–266, https://doi.org/10.1130/G24402A.1

[93] KAGAN, Y.Y., Universality of the seismic-moment-frequency relation, Pure Appl. Geophys. **155** (1999) 537–573, https://doi.org/10.1007/s000240050277

[94] KAGAN, Y.Y., Seismic moment distribution revisited: II. Moment conservation principle, Geophys. J. Int. **149** 3 (2002a) 731–754, https://doi.org/10.1046/j.1365-246X.2002.01671.x

[95] KAGAN, Y.Y., Seismic moment distribution revisited: I. Statistical results, Geophys. J. Int. **148** 3 (2002b) 520–541, https://doi.org/10.1046/j.1365-246x.2002.01594.x

[96] BERRYMAN, K., et al., The GEM Faulted Earth Subduction Characterisation Project, Version 1.0, GEM Faulted Earth Project, Global Earthquake Model (2013), https://www.goosocean.org/index.php?option=com_oe&task=viewDocumentRecord&docID=11672

[97] BIRD, P., KAGAN Y.Y., Plate-tectonic analysis of shallow seismicity: Apparent boundary width, beta-value, corner magnitude, coupled lithosphere thickness, and coupling in 7 tectonic settings, Bull. Seismol. Soc. Am. **94** (2004) 2380–2399, https://doi.org/10.1785/0120030107

[98] WESNOUSKY, S.G., The Gutenberg-Richter or characteristic earthquake distribution, which is it? Bull. Seismol. Soc. Am. **84** 6 (1994) 1940–1959, https://doi.org/10.1785/BSSA0840061940

[99] GOLDFINGER, C., et al., Turbidite event history— Methods and implications for Holocene paleoseismicity of the Cascadia Subduction Zone, U.S. Geol. Surv., Prof. Pap. **1661-F** (2012) 184, https://doi.org/10.3133/pp1661F

[100] ISHIBE, T., SHIMAZAKI, K., Characteristic earthquake model and seismicity around Late Quaternary active faults in Japan, Bull. Seismol. Soc. Am. **102** 3 (2012) 1041–1058, https://doi.org/10.1785/0120100250

[101] MATTHEWS, M.V., ELLSWORTH, W.L., REASENBERG P.A., A Brownian model for recurrent earthquakes, Bull. Seismol. Soc. Am. **92** 6 (2002) 2233–2250, https://doi.org/10.1785/0120010267

[102] DIDENKULOVA, I.I., PELINOVSKY, E.N., Phenomena similar to tsunami in Russian internal basins, Russ. J. Earth Sci. **8** (2006), https://doi.org/10.2205/2006ES000211

[103] VARNES, D.J., “Slope movement types and processes”, Landslides: Analysis and Control (SCHUSTER, R.L., KRIZEK R.J., Eds), National Academy of Sciences, Washington, DC (1978) 11–33, https://onlinepubs.trb.org/onlinepubs/sr/sr176/176-002.pdf

[104] TEN BRINK, U.S., CHAYTOR, J.D., GEIST, E.L., BROTHERS, D.S., ANDREWS, B.D., Assessment of tsunami hazard to the US Atlantic margin, Mar. Geol. **353** (2014) 31–54, https://doi.org/10.1016/j.margeo.2014.02.011

[105] IMRAN, J., PARKER, G., LOCAT, J., LEE, H., 1-D numerical model of muddy subaqueous and subaerial debris flows, J. Hydraul. Eng. **127** 11 (2001) 959–958, https://doi.org/10.1061/(ASCE)0733-9429(2001)127:11(959)

[106] IMRAN, J., HARFF, P., PARKER, G., A numerical model of submarine debris flow with graphical user interface, Comput. Geosci. **27** 6 (2001) 717–729, https://doi.org/10.1016/S0098-3004(00)00124-2

[107] LYNETT, P., LIU, P.L.-F., A numerical study of submarine-landslide-generated waves and run-up, Proc. R. Soc. London, Ser. A Math. Phys. Sci. **458** (2002) 2885–2910, https://doi.org/10.1098/rspa.2002.0973

[108] LYNETT, P., LIU, P.L.-F., A numerical study of run-up generated by three-dimensional landslides, J. Geophys. Res. **110** C3 (2005) C03006, https://doi.org/10.1029/2004JC002443

[109] ENET, F., GRILLI, S.T., Experimental study of tsunami generation by three-dimensional rigid underwater landslides, J. Waterw. Port Coast. Ocean Eng. **133** 6 (2007) 442–454, https://doi.org/10.1061/(ASCE)0733-950X(2007)133:6(442)

[110] DI RISIO, M., et al., Landslide-generated tsunamis runup at the coast of a conical island: New physical model experiments, J. Geophys. Res. **114** C1 (2009) C01009, https://doi.org/10.1029/2008JC004858

[111] LIU, P.L.-F., WU, T.-R., RAICHLEN, F., SYNOLAKIS, C.E., BORRERO, J.C., Runup and rundown generated by three-dimensional sliding masses, J. Fluid Mech. **536** 1 (2005) 107–144, https://doi.org/10.1017/S0022112005004799

[112] LOCAT, J., LEE, H.J., LOCAT, P., IMRAN, J., Numerical analysis of the mobility of the Palos Verdes debris avalanche, California, and its implication for the generation of tsunamis, Mar. Geol. **203** 3–4 (2004) 269–280, https://doi.org/10.1016/S0025-3227(03)00310-4

[113] LOCAT, J., et al., Geomorphology, stability and mobility of the Currituck slide, Mar. Geol. **264** 1–2 (2009) 28–40, https://doi.org/10.1016/j.margeo.2008.12.005

[114] CHAYTOR, J., TEN BRINK, U.S., SOLOW, A.R., ANDREWS, B.D., Size distribution of submarine landslides along the US Atlantic margin, Mar. Geol. **264** 1–2 (2009) 16–27,
https://doi.org/10.1016/j.margeo.2008.08.007

[115] TEN BRINK, U.S., GEIST, E.L., ANDREWS, B.D., Size distribution of submarine landslides and its implication to tsunami hazard in Puerto Rico, Geophys. Res. Lett. **33** L11 (2006) L11307,
https://doi.org/10.1029/2006GL026125

[116] NORMARK, W.R., MCGANN, M., SLITER, R.W., Age of Palos Verdes submarine debris avalanche, Southern California, Mar. Geol. **203** 3–4 (2004) 247–259,
https://doi.org/10.1016/S0025-3227(03)00308-6

[117] GEIST, E.L., PARSONS, T., "Estimating the empirical probability of submarine landslide occurrence", Submarine Mass Movements and their Consequences IV, (MOSHER, D.C., et al., Eds), Springer, Heidelberg (2010) 377–386,
https://doi.org/10.1007/978-90-481-3071-9_31

[118] TEN BRINK, U.S., LEE, H.J., GEIST, E.L., TWICHELL, D.C., Assessment of tsunami hazard to the US East Coast using relationships between submarine landslides and earthquakes, Mar. Geol. **264** (2009) 65–73,
https://doi.org/10.1016/j.margeo.2008.05.011

[119] WARD, S.N., DAY, S., Cumbre Vieja volcano — Potential collapse and tsunami at La Palma, Canary Islands, Geophys. Res. Lett. **28** 17 (2001) 3397–3400,
https://doi.org/10.1029/2001GL013110

[120] NOMANBHOY, N., SATAKE, K., Generation mechanism of tsunamis from the 1883 Krakatau eruption, Geophys. Res. Lett. **22** 4 (1995) 509–512,
https://doi.org/10.1029/94GL03219

[121] POWER, W., DOWNES, G., "Tsunami hazard assessment", Volcanic and Tectonic Hazard Assessment for Nuclear Facilities (CONNOR, C.B., CHAPMAN, N.A., CONNOR, L.J., Eds), Cambridge University Press, Cambridge (2009) 276–306.

[122] OEHLER, J.F., LABAZUY, P., LÉNAT, J.F., Recurrence of major flank landslides during the last 2-Ma–history of Réunion Island, Bull. Volcanol. **66** 7 (2004) 585–598,
http://dx.doi.org/10.1007/s00445-004-0341-2

[123] MOORE, J.G., BRYAN, W.B., LUDWIG, K.R., Chaotic deposition by a giant wave, Molokai, Hawaii, Geol. Soc. Am., Bull. **106** 7 (1994) 962–967,
https://doi.org/10.1130/0016-7606(1994)106%3C0962:CDBAGW%3E2.3.CO;2

[124] MASSON, D.G., et al., Slope failures on the flanks of the western Canary Islands, Earth-Sci. Rev. **57** (2002) 1–35,
https://doi.org/10.1016/S0012-8252(01)00069-1

[125] LØVHOLT, F., PEDERSEN, G., GISLER, G., Oceanic propagation of a potential tsunami from the La Palma Island, J. Geophys. Res. **113** C9 (2008) C09026,
https://doi.org/10.1029/2007JC004603

[126] WAYTHOMAS, C.F., WATTS, P., SHI, F., KIRBY, J.T., Pacific Basin tsunami hazards associated with mass flows in the Aleutian arc of Alaska, Quat. Sci. Rev. **28** 11–12 (2009) 1006–1019, https://doi.org/10.1016/j.quascirev.2009.02.019

[127] GIACHETTI, T., PARIS, R., KELFOUN, K., PEREZ-TORRADO, F.J., Numerical modelling of the tsunami triggered by the Güímar debris avalanche, Tenerife (Canary Islands): Comparison with field-based data, Mar. Geol. **284** 1–4 (2011) 189–202, https://doi.org/10.1016/j.margeo.2011.03.018

[128] BARDET, J.P., SYNOLAKIS, C.E., DAVIES, H.L., IMAMURA, F., OKAL, E.A., Landslide tsunamis: Recent findings and research directions, Pure Appl. Geophys. **160** 10 (2003) 1793–1809, https://doi.org/10.1007/978-3-0348-7995-8_1

[129] HARBITZ, C.B., LØVHOLT, F., PEDERSEN, G., MASSON, D.G., Mechanisms of tsunami generation by submarine landslides: a short review, Nor. J. Geol. **86** (2006) 255–264, https://njg.geologi.no/images/NJG_articles/60713_NGT_no_3_06_11.pdf

[130] HELLER, V., HAGER, W.H., Impulse product parameter in landslide-generated impulse waves, J. Waterw. Port Coast. Ocean Eng. **136** 3 (2010) 145–155, https://doi.org/10.1061/(ASCE)WW.1943-5460.0000037

[131] WATT, S.F., et al., Combinations of volcanic-flank and seafloor-sediment failure offshore Montserrat, and their implications for tsunami generation, Earth Planet. Sci. Lett. **319–320** (2012) 228–240, https://doi.org/10.1016/j.epsl.2011.11.032

[132] TINTI, S., PAGNONI, G., ZANIBONI, F., The landslides and tsunamis of the 30th of December 2002 in Stromboli analysed through numerical simulations, Bull. Volcanol. **68** 5 (2006) 462–479, https://doi.org/10.1007/s00445-005-0022-9

[133] LATTER, J.H., Tsunamis of volcanic origin: Summary of causes with particular reference to Krakatoa, 1883, Bull. Volcanol. **44** 3 (1981) 467–490, https://doi.org/10.1007/BF02600578

[134] DE LANGE, W.P., PRASETYA, G.S., HEALY, T.R., Modelling of tsunamis generated by pyroclastic flows (Ignimbrites), Nat. Hazards **24** 3 (2001) 251–266, https://doi.org/10.1023/A:1012056920155

[135] PARIS, R., et al., Volcanic tsunami: A review of source mechanisms, past events and hazards in Southeast Asia (Indonesia, Philippines, Papua New Guinea), Nat. Hazards **70** 1 (2014) 447–470, https://doi.org/10.1007/s11069-013-0822-8

[136] SIMKIN, T., FISKE, R.S. (Eds), Krakatau, 1883 — The Volcanic Eruption and its Effects, Smithsonian Institution Press, Washington, DC (1983).

[137] SELF, S., RAMPINO, M.R., The 1883 eruption of Krakatau, Nature **294** (1981) 699–704, https://doi.org/10.1038/294699a0

[138] WAYTHOMAS, C.F., WATTS, P., Numerical simulation of tsunami generation by pyroclastic flow at Aniakchak Volcano, Alaska, Geophys. Res. Lett. **30** 14 (2003), https://doi.org/10.1029/2003GL017220

[139] MAENO, F., IMAMURA, F., TANIGUCHI, H., Numerical simulation of tsunamis generated by caldera collapse during the 7.3 ka Kikai eruption, Kyushu, Japan, Earth Planets Space **58** 8 (2006) 1013–1024, https://doi.org/10.1186/BF03352606

[140] MAENO, F., IMAMURA, F., Tsunami generation by a rapid entrance of pyroclastic flow into the sea during the 1883 Krakatau eruption, Indonesia, J. Geophys. Res. **116** B9 (2011) B09205, https://doi.org/10.1029/2011JB008253

[141] VOIGHT, B., GLICKEN, H., JANDA, R.J., DOUGLASS, P.M., "Catastrophic rockslide avalanche of May 18", The 1980 Eruptions of Mount St. Helens, Washington (LIPMAN, P.W., MULLINEAUX, D.R., Eds), US Geol. Surv. Prof. Pap. 1250, Reston, VA (1981) 347–377, https://pubs.usgs.gov/pp/1250/report.pdf

[142] VOIGHT, B., JANDA, R., GLICKEN, H., DOUGLAS, P.M., Nature and mechanics of the Mount St. Helens rockslide-avalanche of 18 May 1980, Géotech. **33** 10 (1983) 243–273, http://dx.doi.org/10.1680/geot.1983.33.3.243

[143] WARD, S.N., DAY, S., Ritter Island volcano — Lateral collapse and the tsunami of 1888, Geophys. J. Int. **154** 3 (2003) 891–902, https://doi.org/10.1046/j.1365-246X.2003.02016.x

[144] HEINRICH, P., et al., Numerical simulation of the December 1997 debris avalanche in Montserrat, Lesser Antilles, Geophys. Res. Lett. **28** 13 (2001) 2529–2532, https://doi.org/10.1029/2001GL012968

[145] AIDA, I., Numerical experiments of the tsunamis associated with the collapse of Mt. Mayuyama in 1792, J. Seismolog. Soc. Jpn. **28** (1975) 449–460 (in Japanese with English abstract), https://doi.org/10.4294/zisin1948.28.4_449

[146] INTERNATIONAL ATOMIC ENERGY AGENCY, Volcanic Hazard Assessments for Nuclear Installations: Methods and Examples in Site Evaluation, IAEA-TECDOC-1795, IAEA, Vienna (2016), https://www-pub.iaea.org/MTCD/Publications/PDF/TE1795web.pdf

[147] STOOPES, G., SHERIDAN, M.F., Giant debris avalanches from the Colima Volcanic Complex, Mexico: Implications for long-runout landslides (>100 km) and hazard assessment, Geology **20** 4 (1992) 299–302, https://doi.org/10.1130/0091-7613(1992)020%3C0299:GDAFTC%3E2.3.CO;2

[148] UI, T., YAMAMOTO, H., SUZUKI-KAMATA, K., Characterization of debris avalanche deposits in Japan, J. Volcanol. Geotherm. Res. **29** (1986) 231–243, https://doi.org/10.1016/0377-0273(86)90046-6

[149] KELFOUN, K., GIACHETTI, T., LABAZUY, P., Landslide-generated tsunamis at Réunion Island, J. Geophys. Res. **115** F4 (2010) F04012, https://doi.org/10.1029/2009JF001381

[150] YOKOYAMA, I., A geophysical interpretation of the 1883 Krakatau eruption, J. Volcanol. Geotherm. Res. **9** (1981) 359–378, https://doi.org/10.1016/0377-0273(81)90044-5

[151] TORSVIK, T., et al., Numerical simulation of a tsunami event during the 1996 volcanic eruption in Karymskoye lake, Kamchatka, Russia, Nat. Hazards Earth Syst. Sci. **10** (2010) 2359–2369, https://doi.org/10.5194/nhess-10-2359-2010

[152] SATO, H., TANIGUCHI, H., Relationship between crater size and ejecta volume of recent magmatic and phreato-magmatic eruptions: Implications for energy partitioning, Geophys. Res. Lett. **24** 3 (1997) 205–208, https://doi.org/10.1029/96GL04004

[153] MAENO, F., IMAMURA, F., Numerical investigations of tsunamis generated by pyroclastic flows from the Kikai caldera, Japan, Geophys. Res. Lett. **34** L23 (2007) L23303, https://doi.org/10.1029/2007GL031222

[154] WALDER, J.S., WATTS, P., SORENSEN, O.E., JANSSEN, K., Tsunamis generated by subaerial mass flows, J. Geophys. Res. **108** B5 (2003) EPM 2-1, https://doi.org/10.1029/2001JB000707

[155] LE MEHAUTE, B., WANG, S., Water Waves Generated by Underwater Explosions, Technical Report DNA-TR-94-128, Defense Nuclear Agency, Alexandria, VA (1996), https://apps.dtic.mil/sti/citations/tr/ADA304244

[156] LE MEHAUTE, B., KHANGAONKAR, T., Generation and Propagation of Explosion Generated Waves in Shallow Water, Technical Report DNA-TR-92-40, Defense Nuclear Agency, Alexandria, VA (1992).

[157] BEGET, J.E., KOWALIK, Z., Confirmation and calibration of computer modeling of tsunami produced by Augustine volcano, Alaska, Sci. Tsunami Hazards **24** 4 (2006) 257–266.

[158] OZEROV, A., ISPOLATOV, I., LEES, J., Modeling Strombolian eruptions of Karymsky volcano, Kamchatka, Russia, J. Volcanol. Geotherm. Res. **122** 3 (2003) 265–280, https://doi.org/10.1016/S0377-0273(02)00506-1

[159] VERBEEK, R.D.M., The Krakatoa eruption, Nature **30** 757 (1884) 10–15, https://doi.org/10.1038/030010a0

[160] SYMONS, G.J. (Ed.), The Eruption of Krakatoa and Subsequent Phenomena, Report of the Krakatoa Committee of the Royal Society of London, Trubner, London (1888).

[161] YOKOYAMA, I., A scenario of the 1883 Krakatau tsunami, J. Volcanol. Geotherm. Res. **34** (1987) 123–132, https://doi.org/10.1016/0377-0273(87)90097-7

[162] CHOI, B.H., PELINOVSKY, E., KIM, K.O., LEE, J.S., Simulation of the trans-oceanic tsunami propagation due to the 1883 Krakatau volcanic eruption, Nat. Hazards Earth Syst. Sci. **3** 5 (2003) 321–332, https://doi.org/10.5194/nhess-3-321-2003

[163] INTERNATIONAL ATOMIC ENERGY AGENCY, Volcanic Hazards in Site Evaluation for Nuclear Installations, IAEA Safety Standards Series No. SSG-21, IAEA, Vienna (2012), https://www.iaea.org/publications/8774/volcanic-hazards-in-site-evaluation-for-nuclear-installations

[164] FUJIMA K., MASAMURA, K., GOTO, C., Development of the 2D/3D hybrid model for tsunami numerical simulation, Coast. Eng. J. **44** 4 (2002) 373–397, https://doi.org/10.1142/S0578563402000615

[165] KIM, D.C., KIM, K.O., PELINOVSKY, E., DIDENKULOVA, I., CHOI, B.H., Three-dimensional tsunami runup simulation for the port of Koborinai on the Sanriku coast of Japan, J. Coast. Res. Special Issue **65** (2013) 266–271, https://doi.org/10.2112/SI65-046.1

[166] IWASE, H., IMAMURA, F., "A new tsunami numerical simulation with Boussinesq-type equations applied for the 1983 Nihonkai-Chubu Earthquake tsunami", Proc. APAC 2003, Makuhari, Japan (2004) 12–13.

[167] BURWELL, D., TOLKOVA, E., CHAWLA, A., Diffusion and dispersion characterization of a numerical tsunami model, Ocean Model. **19** (2007) 10–30, https://doi.org/10.1016/j.ocemod.2007.05.003

[168] ZHOU, H., WEI, Y., TITOV, V.V., Dispersive modelling of the 2009 Samoa tsunami, Geophys. Res. Lett. **39** 16 (2012) L16603, https://doi.org/10.1029/2012GL053068

[169] SHUTO, N., SUZUKI, T., HASEGAWA, K., INAGAKI, K., A study of numerical techniques on the tsunami propagation and run-up, Sci. Tsunami Hazards **4** 2 (1986) 111–124.

[170] IUGG/IOC TIME PROJECT, IOC Manuals and Guides No. 35, UNESCO, Paris (1997).

[171] IMAMURA, F., "Tsunami modeling: calculating inundation and hazard maps", The Sea: Ideas and Observations on Progress in the Study of the Seas: Tsunamis, 1st edn, Vol. 15 (BERNARD, E.N., ROBINSON, A.R., Eds), Harvard University Press, Cambridge, MA (2009) 321–332.

[172] AIDA, I., Simulations of Large Tsunamis Occurring in the Past off the Coast of the Sanriku District, Bulletin of the Earthquake Research Institute, University of Tokyo, Vol. 52(1), (1977) 71-101 (in Japanese), https://repository.dl.itc.u-tokyo.ac.jp/?action=repository_action_common_download&item_id=33218&item_no=1&attribute_id=19&file_no=1

[173] HOUSTON, J., BUTLER, H.L., Numerical Model for Tsunami Inundation, Technical Report, Hydraulics Laboratory, U.S. Army Engineer Waterways Experiment Station, Vicksburg, Mississippi (1979).

[174] HIBBERD, S., PERGRINE, D.H., Surf and run-up on a beach: a uniform bore, J. Fluid Mech. **95** (1979) 323-345.

[175] IWASAKI, T., MANO, A., "Two-dimensional numerical simulation of tsunami runups in the Eulerian description", Proc. 26th Conf. on Coastal Engineering, JSCE, Tokyo (1979) 70–72 (in Japanese),
http://library.jsce.or.jp/jsce/open/00008/1979/26-0070.pdf

[176] IMAMURA, F., "Review of tsunami simulation with a finite difference method", Long-Wave Runup Models (Yeh, H., Liu, P., Synolakis, C., Eds), World Scientific, (1996) 25-42.

[177] HOM-MA, H., Coefficient of flow volume on low overflow weir, Doboku Gakkai Ronbun Hokokushu **26** 6 (1940) 635–645 (in Japanese).

[178] GOTO, C., SATO, K., Development of Tsunami Numerical Simulation System for Sanriku Coast in Japan, Report of the Port and Harbour Research Institute, Vol. 32, No. 2 (1993) 3–40 (in Japanese).

[179] KOTANI, M., IMAMURA, F., SHUTO, N., Tsunami run-up simulation and damage estimation by using GIS, Proc. Coast. Eng. **45** (1998) 356–360 (in Japanese).

[180] TANAKA, H., Development of Mathematical Models for Tsunami Behavior on Coastal Zones, Abiko Research Laboratory Report of Central Research Institute of Electric Power Industry, No. 385017 (1985) 46 (in Japanese).

[181] DENGLER, L.A., USLU, B., BARBEROPOULOU, A., BORRERO, J.C., SYNOLAKIS, C.E., The vulnerability of Crescent City, California, to tsunamis generated by earthquakes in the Kuril Islands region of the Northwestern Pacific, Seismol. Res. Lett. **79** 5 (2008) 608–619,
https://doi.org/10.1785/gssrl.79.5.608

[182] INTERNATIONAL ATOMIC ENERGY AGENCY, Benchmark Analysis of Numerical Models for Tsunami Simulation, IAEA-TECDOC-1973, IAEA, Vienna (2022),
https://www.iaea.org/publications/14923/benchmark-analysis-of-numerical-models-for-tsunami-simulation

[183] SYNOLAKIS, C.E., et al., Validation and verification of tsunami numerical models, Pure Appl. Geophys. **165** 11–12 (2008) 2197–2228,
https://doi.org/10.1007/s00024-004-0427-y

[184] NATIONAL OCEANIC AND ATMOSPHERIC ADMINISTRATION, Standards, Criteria, and Procedures for NOAA Evaluation of Tsunami Numerical Models, NOAA Technical Memorandum OAR PMEL-135, Pacific Marine Environmental Laboratory, Seattle, WA,
http://www.pmel.noaa.gov/pubs/PDF/syno3053/syno3053.pdf

[185] NATIONAL OCEANIC AND ATMOSPHERIC ADMINISTRATION, Proceedings and Results of the 2011 NTHMP Model Benchmarking Workshop, National Tsunami Hazard Mitigation Program, NOAA Special Report, US Dept. of Commerce/NOAA/NTHMP, Boulder, July 2012, 436 pp.

[186] ANNAKA, T., SATAKE, K., SAKAKIYAMA, T., YANAGISAWA, K., SHUTO, N., Logic-tree approach for probabilistic tsunami hazard analysis and its applications to the Japanese coasts, Pure Appl. Geophys. **164** 2–3 (2007) 577–592,
https://doi.org/10.1007/s00024-006-0174-3

[187] GEIST, E.L., Complex earthquake rupture and local tsunamis, J. Geophys. Res., **107** B5 (2002), ESE 2-1–ESE 2-15,
https://doi.org/10.1029/2000JB000139

[188] GEIST, E.L., LYNETT, P.J., Source processes for the probabilistic assessment of tsunami hazards, Oceanogr. **27** 2 (2014) 86–93,
https://doi.org/10.5670/oceanog.2014.43

[189] BUDNITZ, R.J., et al., Recommendations for Probabilistic Seismic Hazard Analysis: Guidance on Uncertainty and Use of Experts, Vol. 2, NUREG/CR-6372/UCRL-ID-122160, NRC/LLNL, Washington, DC/LLNL, Livermore, CA (1997).

[190] HANKS, T.C., ABRAHAMSON, N.A., BOORE, D.M., COPPERSMITH, K.J., KNEPPRATH, N.E., Implementation of the SSHAC guidelines for Level 3 and 4 PSHAs — experience gained from actual applications, U.S. Geol. Surv. (2009) 66,
https://doi.org/10.3133/ofr20091093

[191] RIKITAKE, T., AIDA, I., Tsunami hazard probability in Japan, Bull. Seismol. Soc. Am. **78** 3 (1988)1268–1278,
https://doi.org/10.1785/BSSA0780031268

[192] GEIST, E.L., PARSONS, T., Probabilistic analysis of tsunami hazards, Nat. Hazards **37** (2006) 277–314,
https://doi.org/10.1007/s11069-005-4646-z

[193] GEIST, E.L., PARSONS, T., TEN BRINK, U.S., LEE, H.J., "Tsunami probability", The Sea: Ideas and Observations on Progress in the Study of the Seas: Tsunamis, 1st edn, Vol. 15 (BERNARD, E.N., ROBINSON, A.R., Eds) Harvard University Press, Cambridge, MA (2009).

[194] FLINN, E.A., ENGDAHL, E.R., HILL, A.R., Seismic and geographical regionalization, Bull. Seismol. Soc. Am. **64** (1974) 771–992,
https://doi.org/10.1785/BSSA0643-20771

[195] McGUIRE, R.K., Probabilistic seismic hazard analysis and design earthquakes: Closing the loop, Bull. Seismol. Soc. Am. **85** 5 (1995) 1275–1284,
https://doi.org/10.1785/BSSA0850051275

[196] BAZZURRO, P., CORNELL, C.A., Disaggregation of seismic hazard, Bull. Seismol. Soc. Am. **89** 2 (1999) 501–520,
https://doi.org/10.1785/BSSA0890020501

[197] PETERSEN, M.D., CAO, T., CAMPBELL, K.W., FRANKEL, A.D., Time-independent and time-dependent seismic hazard assessment for the State of California: Uniform California earthquake rupture forecast model 1.0, Seismol. Res. Lett. **78** 1 (2007) 99–109,
https://doi.org/10.1785/gssrl.78.1.99

[198] ATOMIC ENERGY SOCIETY OF JAPAN, Implementation Standard Concerning the Tsunami Probabilistic Risk Assessment of Nuclear Power Plants: 2011, AESJ-SC-RK004E: 2011, AESJ, Tokyo (2012).

[199] THIO, H.K., SOMERVILLE, P., POLET, J., Probabilistic tsunami hazard in California, PEER 2010/108, Pacific Earthquake Engineering Research Center, Berkeley, CA (2010),
https://peer.berkeley.edu/publications/2010-108

[200] TSUNAMI PILOT STUDY WORKING GROUP, Seaside, Oregon Tsunami Pilot Study — Modernization of FEMA Flood Hazard Maps, Open-File Report 2006-1234, US Department of the Interior/US Geological Survey (2006),
http://pubs.usgs.gov/of/2006/1234/of2006-1234.pdf

[201] GONZÁLEZ, F.I., et al. Probabilistic tsunami hazard assessment at Seaside, Oregon for near- and far-field seismic sources, J. Geophys. Res. **114** (2009) C11023,
https://doi.org/10.1029/2008JC005132

[202] PARSONS, T., GEIST, E.L., Tsunami probability in the Caribbean Region, Pure Appl. Geophys. **165** (2009) 2089–2116,
https://doi.org/10.1007/s00024-008-0416-7

[203] YEH, H., SHUTO, N., Tsunami forces and effects on structures, J. Disaster Res. **4** 6 (2009) 375–376,
http://doi.org/10.20965/jdr.2009.p0375

[204] ARIKAWA, T., Structural behavior under impulsive tsunami loading, J. Disaster Res. **4** 6 (2009) 377–381,
http://doi.org/10.20965/jdr.2009.p0377

[205] OSHNACK, M.E., AGUINIGHA, F., COX, D., GUPTA, R., LINDT, J., Effectiveness of small onshore seawall in reducing forces induced by tsunami bore: Large scale experimental study, J. Disaster Res. **4** 6 (2009) 382–390,
http://doi.org/10.20965/jdr.2009.p0382

[206] ARNASON H., PETROFF, C., YEH, H., Tsunami bore impingement onto a vertical column, J. Disaster Res. **4** 6 (2009) 391–403,
http://doi.org/10.20965/jdr.2009.p0391

[207] FUJIMA, K., ACHMAD F., SHIGIHARA, Y.A., MIZUTANI, N., Estimation of tsunami force acting on rectangular structures, J. Disaster Res. **4** 6 (2009) 404–409,
http://doi.org/10.20965/jdr.2009.p0404

[208] LUKKUNAPRASIT, P., THANASISATHIT, N., YEH, H., Experimental verification of FEMA P646 tsunami loading, J. Disaster Res. **4** 6 (2009) 410–418,
http://doi.org/10.20965/jdr.2009.p0410

[209] FEDERAL EMERGENCY MANAGEMENT AGENCY, Guidelines for Design of Structures for Vertical Evacuation from Tsunamis, 2nd edn, FEMA, Washington, DC (2012) 194 pp,
https://www.atcouncil.org/files/FEMA%20P-646small.pdf

[210] MUTLU SUMER, B., et al., Earthquake-induced liquefaction around marine structures, J. Waterw. Port Coast. Ocean Eng. **133** 1 (2007) 55–82,
https://doi.org/10.1061/(ASCE)0733-950X(2007)133:1(55)

[211] SYNOLAKIS, C.E., “Tsunami and seiche”, Earthquake Engineering Handbook, (CHEN, W.F., SCAWTHORN, C., Eds), CRC Press, New York (2003), chapter 9.

[212] SHUTO, N., The Nihonkai–Chubu earthquake tsunami on the north Akita coast, Proc. Coast. Eng. **28** (1985) 255–264, https://doi.org/10.1080/05785634.1985.11924420

[213] TANIMOTO, K., Field and Laboratory Investigations of the Tsunami Caused by 1983 Nihonkai Chubu Earthquake, Technical Note of the Port and Harbour Research Institute No. 470 (1983) 1–299 (in Japanese), https://www.pari.go.jp/en/1983/11/198311047001.html

[214] KIMURA, K., MIZUNO, Y., TSURUYA, H., NAKAGAWA, Y., "Breakwater damage in Okushiri port due to the Hokkaido Nansei-Oki earthquake tsunami", Proc. 25th Int. Conf. on Coastal Engineering, ASCE (1996) 4676–4688, https://doi.org/10.1061/9780784402429.364

[215] KATO, F., INAGAKI, S., FUKUHAMA, M., "Wave force on coastal dike due to tsunami", Proc. 30th Int. Conf. on Coastal Engineering, ASCE (2006) 5150–5161, https://www.pwri.go.jp/eng/ujnr/joint/37/paper/13kato.pdf

[216] IKENO, M., et al., "Effects of soliton fission and wave breaking on tsunami force acting on breakwater", Proc. 30th Int. Conf. on Coastal Engineering, ASCE (2006) 5162–5174, https://doi.org/10.1142/9789812709554_0432

[217] TANIMOTO, K., TSURUYA, K., NAKANO, S., Tsunami force of Nihonkai-Chubu earthquake in 1983 and cause of revetment damage, Proc. Coast. Eng. **31** (1984) 257–261 (in Japanese), https://doi.org/10.2208/proce1970.31.257

[218] MINISTRY OF AGRICULTURE, FORESTRY AND FISHERIES, JAPAN, The Basic Solution for Fishery Harbours as Measures for Earthquake and Tsunami after the 2011 Great East Japan Earthquake and Tsunami, MAFF, Tokyo (2012) 36 (in Japanese), https://www.jfa.maff.go.jp/j/gyoko_gyozyo/g_gideline/attach/pdf/index-64.pdf

[219] IIZUKA, H., MATSUTOMI, H., Damage due to inundation flow of tsunami, Proc. Coast. Eng. **47** (2000) 381–385 (in Japanese), https://doi.org/10.2208/proce1989.47.381

[220] ASAKURA, R., et al., "The tsunami wave force acting on land structures", Proc. 28th Int. Conf. on Coastal Engineering, ASCE (2002) 1191–1202, https://doi.org/10.1142/9789812791306_0101

[221] SAKAKIYAMA, T., Tsunami inundation flow and tsunami pressure on structures, Proc. Coast. Eng. **2** (2012) 771–775 (in Japanese). https://doi.org/10.2208/kaigan.68.I_771

[222] MATSUTOMI, H., OKAMOTO, K., HARADA, K., Inundation flow velocity of tsunami on land and its practical use, Proc. Coast. Eng. **1** 32 (2010), https://icce-ojs-tamu.tdl.org/icce/index.php/icce/article/view/1141/pdf_215

[223] FIRE AND DISASTER MANAGEMENT AGENCY, Research Report on Protection of Facilities for Dangerous Material from Tsunami and Inundation, FDMA (2009) (in Japanese),
https://www.fdma.go.jp/pressrelease/houdou/items/h21/2105/210526-1houdou/02_stunami_houkokusyo.pdf

[224] FUJII, N., et al., Hydraulic experiment on flow and topography change in harbor due to tsunami and its numerical simulation, Dobuku Gakkai-Shi **65** 1 (2009) 291–295 (in Japanese),
https://doi.org/10.2208/kaigan.65.291

[225] KIHARA, N., FUJII, N., MATSUYAMA, M., Three-dimensional sediment transport processes on tsunami-induced topography changes in a harbor, Earth Planets Space **65** (2012) 787–797,
http://dx.doi.org/10.5047/eps.2011.05.036

[226] TAKAHASHI, T., SHUTO, N., IMAMURA, F., ASAI, D., “Modeling sediment transport due to tsunamis with exchange rate between bed load and suspended load layer”, Proc. 27th Int. Conf. on Coastal Engineering, ASCE, Vol. 2 (2000) 1508–1519,
https://doi.org/10.1061/40549(276)117

[227] IKENO, M., YOSHII, T., MATSUYAMA, M., FUJII, N., Estimation of pickup rate of suspended sand by tsunami experiment and proposal of pickup rate formula, Doboku Gakkai-Shi **65** 1 (2009) 506–510 (in Japanese),
https://doi.org/10.2208/kaigan.65.506

[228] YOSHII, T., IKENO, M., MATSUYAMA, M., FUJII, N., Pick-up rate of suspended sand due to tsunami, Coast. Eng. Proc. **1** (32), sediment.12 (2011),
https://journals.tdl.org/icce/index.php/icce/article/view/1068/pdf_28

[229] ASHIDA, K., MICHIUE, M., Study on hydraulic resistance and bed-load transport in alluvial streams, Doboku Gakkai Ronbun Hokokushu **206** (1972) 59–69 (in Japanese),
https://doi.org/10.2208/jscej1969.1972.206_59

[230] GOTO, C., “Simulation models of timber and oil spread due to tsunamis”, Proc. 2nd UJNR Tsunami Workshop, Honolulu, Hawaii, 5–6 November 1990, National Geophysical Records Documentation No. 24, Boulder, CO, NOAA (1990) 63–66,
https://nehrpsearch.nist.gov/static/files/NIST/PB91193656.pdf

[231] KUMAGAI, K., ODA, K., FUJII, N., “Applicability of simulation model for drift behavior of containers due to tsunami”, Proc. Techno-Ocean 2006, JASNAOE Ocean Engineering Symposium, No. 72 (2006),
https://www.ysk.nilim.go.jp/kakubu/engan/engan/taigai/kumagai_taigai/20061019.pdf

[232] FUJII, N., et al., Numerical simulation of tsunami drifting bodies in a harbour, Proc. Coast. Eng. **52** (2005) 296–300 (in Japanese),
https://doi.org/10.2208/proce1989.52.296

[233] INAGAKI, S., et al., Model experiment and simulative prediction on slipping and drifting of oil storage tank by tsunami, Proc. Coastal Eng. **55** (2008) 276–280 (in Japanese),
https://doi.org/10.2208/proce1989.55.276

Annex

EXAMPLES OF TSUNAMI DATABASES

A–1. NOAA NCEI/WDS DATABASES

A–1.1. Introduction and background information

The National Geophysical Data Center (NGDC) in the United States of America was formerly one of the three national data centres operated by the National Oceanic and Atmospheric Administration's (NOAA) National Environmental Satellite, Data, and Information Service (NESDIS). In 2015, NGDC merged with two other NESDIS data centres (the National Climate Data Center and the National Oceanographic Data Center) to form the National Centers for Environmental Information (NCEI). NCEI's mission is to provide long term scientific data stewardship for the USA's coastal, geophysical and oceanographic data. NCEI also operates the World Data Service (WDS) for Geophysics, which manages global geophysical, seafloor and natural hazards data, including data on tsunamis. The NCEI global tsunami archive incorporates the historical tsunami database, imagery, and raw and processed US coastal tide gauge and tsunameter data, including Deep-ocean Assessment and Reporting of Tsunami (DART) relevant to a tsunami event. Since the majority of tsunamis are generated by earthquakes, NCEI also maintains a significant earthquake database. NCEI provides integrated access to these data via web forms, interfaces and interactive maps.

A–1.2. NCEI/WDS Global Historical Tsunami Database

The NCEI/WDS Global Historical Tsunami Database[1] includes all events regardless of magnitude or intensity. As of 2019, over 2200 source events (not erroneous or seiches) had been recorded from 2000 BCE to the present. Since the written record is fairly complete from 1701 CE onward, Fig. A–1 provides a count of tsunamis for ten-year periods from 1701 to the present. The increase in the numbers is due to the increase in observations and awareness of tsunamis. For example, if today a tsunami generates a wave of a few centimetres that is observed only in the deep ocean at a tsunameter station, it is considered a tsunami and added to the database. Also, the identification of a tsunami signature from tide gauge information requires higher resolution data than is usually collected

[1] https://doi.org/10.7289/V5PN93H7

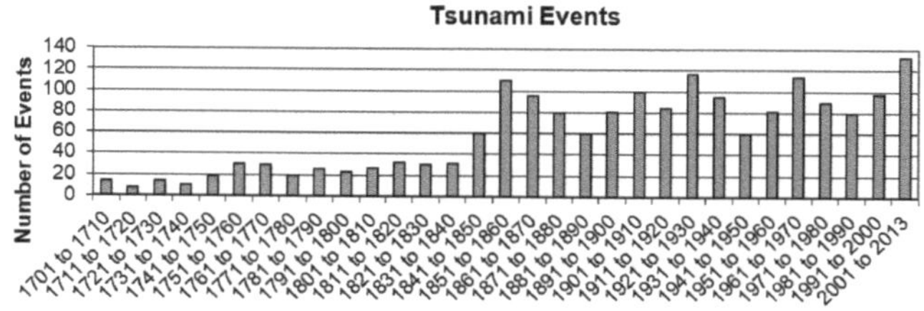

FIG. A–1. Distribution of tsunami events (not erroneous) from 1701 to 2013. Source: NCEI/WDS.

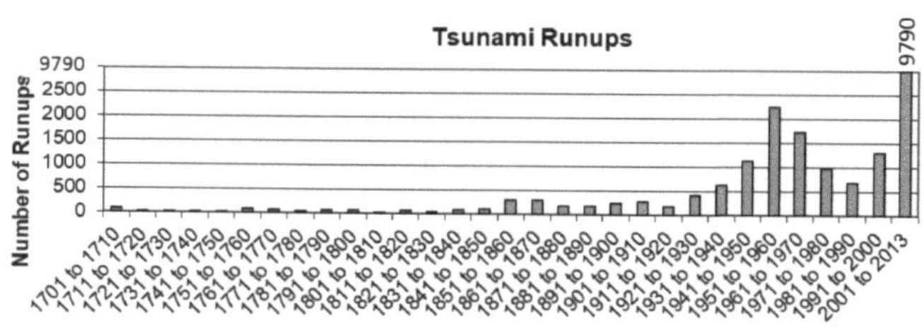

FIG. A–2. Distribution of tsunami runups from 1701 to 2013. Source: NCEI/WDS.

for tide predictions. Since the 2004 Indian Ocean tsunami, many tide stations have been upgraded from a six-minute dissemination of water level data to one minute. Tsunameter or DART and bottom pressure recorder stations, developed in the 1980s and now installed all over the world, are also major tools used to identify tsunamis. These enhancements allow scientists to observe small tsunamis that would have been missed in the past. The global distribution of tsunami events is 70% the Pacific Ocean, 15% the Mediterranean Sea and the Black Sea, 9% the Caribbean Sea and the Atlantic Ocean, and 6% the Indian Ocean.[2]

The tsunami database includes additional information on runups (locations where tsunami waves were observed by eyewitnesses, post-tsunami field surveys, tide gauges or deep ocean sensors). As of 2019, there have been over 27 000 runup observations since 2000 BCE. Figure A–2 provides a count of tsunami runups for ten-year periods from 1701 to 2013. The distribution of tsunami runups is 87% in the Pacific Ocean, 3% Mediterranean Sea and Black Sea, 2% Atlantic Ocean and the East Coast of the USA and Canada, 7% Indian Ocean and 1% Caribbean Sea. The distribution of runups in the last 50–70 years is mostly due to field surveys conducted following large events, such as the 2004 Indian Ocean and 2011 Tohoku, Japan, tsunamis.

[2] https://www.ngdc.noaa.gov/hazard/tsunami-db-intro.html

A–1.3. Source information

Each event or runup in the database is associated with a list of one or more source documents. More than 1700 source documents were used to compile the tsunami event and runup records. The sources include tsunami catalogues, journal articles, field surveys, newspaper articles, instrumental records (tide gauge, deep ocean sensor), the NOAA/National Weather Service Tsunami Warning Centers, the United States Geological Survey National Earthquake Information Center, the Intergovernmental Oceanographic Commission of UNESCO, etc. In some cases, different sources provide differing information for a particular record. NCEI records this information in an event comment field.

A–1.4. Tsunami event and runup tables

The NCEI/WDS Tsunami Event Database includes the date, time and location of the source event, type of source (e.g. earthquake, volcano, landslide, explosion, unknown), the magnitude of the source, event validity, maximum wave height, tsunami magnitude and intensity, total number of runups, and total number of fatalities, injuries, houses damaged and dollar damage from the tsunami, with the source event tabulated separately. If the source event was an earthquake or volcanic eruption, it is linked to the corresponding database that provides additional information on the source. The tsunami runup table includes the location name, latitude, longitude, country, type of measurement, calculated and observed arrival times, distance from the source, maximum inundation distance, maximum wave height, time of maximum wave height, first motion of the wave, wave period, and total number of fatalities, injuries, houses damaged and dollar damage at the specific runup location. Table A–1 lists 15 tsunami events for which the number of runup data exceeds 200.

A–1.5. Tsunami event validities

A validity score is assigned to each tsunami source event, ranging from –1 for erroneous entries, 0 for seiches, 1 to 2 for unconfirmed tsunamis and 3 to 4 for confirmed tsunamis. The validity of tsunami events and runups is based on several factors. For example, tsunamis recorded on tide gauges generated by earthquakes that were recorded on seismographs are assigned a high validity of 4 in the database.

Historical events that occurred before the invention of the seismograph or tide gauge must be evaluated differently. If the event had significant effects, such as deaths and damage, or was observed in many locations, it is also considered a high validity event. For example, a tsunami generated by an earthquake in Chile

TABLE A–1. TSUNAMI EVENTS THAT GENERATED OVER 200 RUNUPS

Date	Source location	Earthquake magnitude	Maximum runup (m)	Number of runups
11 Mar. 2011	Tohoku, Japan	9.0	38.90	5776
26 Dec. 2004	Sumatra, Indonesia	9.1	50.90	1058
22 May 1960	Central Chile	9.5	25.00	1050
27 Feb. 2010	Southern Chile	8.8	29.00	597
29 Sep. 2009	Samoa Islands	8.0	22.35	579
1 Apr. 1946	Unimak Island, Alaska, USA	8.1	35.05	511
28 Mar. 1964	Prince William Sound, Alaska, USA	9.2	67.10	394
9 Mar. 1957	Andreanof Islands, Alaska, USA	8.6	22.80	323
16 May 1968	Honshu Island, Japan	8.2	6.00	306
20 Dec. 1946	Honshu Island, Japan	8.1	6.60	298
2 Mar. 1933	Sanriku, Japan	8.4	29.00	295
4 Nov. 1952	Kamchatka, Russia	9.0	18.00	290
26 May 1983	Noshiro, Japan	7.8	14.93	227
1 Apr. 2007	Solomon Islands	8.1	12.10	224
4 Mar. 1952	Hokkaido, Japan	8.1	6.50	219

(source: NCEI/WDS, as of 2013)

TABLE A–2. DISTRIBUTION OF NCEI/WDS TSUNAMI EVENTS BY EVENT VALIDITY (as of 2013)

Tsunami event	
Count (as of 2013)	Validity
181	–1
94	0
445	1
512	2
357	3
837	4

that was observed in Hawaii and California would be assigned a high validity. The number of reliable and independent sources that list a historical event also affects the validity. Historical tsunami events generated by earthquakes are cross-checked with regional and local earthquake catalogues. If the tsunami was reported to have been generated by an earthquake, but there are no listings in the earthquake catalogues, the validity is lowered. Tsunami events generated by volcanoes are cross-checked with volcano catalogues.

As of 2013, there were 180 erroneous, 93 seiches, 958 unconfirmed and 1195 confirmed tsunamis in the NCEI database (see Table A–2). Tsunamis with event validities of –1 (erroneous) are left in the database, since they may be listed as valid in other databases or tsunami catalogues. In this way, it is clear to the user that an event has been determined to be erroneous and not just missed.

A–1.6. Tsunami runup validities

The tsunami runups are not assigned a numeric validity, but the validity of the source event applies to the runup. If the runup is questionable or the cause was meteorological, it is flagged. Also, the type of measurement is indicative of the validity. For example, eyewitness reports are usually less reliable than instrumental recordings or post-tsunami field surveys.

A–1.7. NCEI/WDS Global Significant Earthquake and Significant Volcanic Eruption databases

The NCEI/WDS Global Significant Earthquake[3] and Significant Volcanic Eruption[4] databases include events that range in date from 4360 BCE to the present. These databases include all earthquakes and volcanic eruptions that either caused fatalities or moderate damage or generated a tsunami. Earthquakes that were assigned either a magnitude >7.4 or a modified Mercalli intensity >IX and eruptions with a volcanic explosivity index >5 are also included. If the earthquake or eruption generated a tsunami, it is linked to the tsunami event table. As of 2013, the database included 5720 earthquakes and 635 volcanic eruptions.

A–1.8. Data access and delivery

A primary goal of NCEI/WDS is to make the data easily discoverable and integrated (i.e. a user can link from one dataset to related data in another dataset). The historical event data are stored in a spatially enabled relational database management system that facilitates integration and access to these related data tables. The data can be searched from web form interfaces by attribute (date, location, magnitude, etc.) and displayed as tables, reports, interactive maps and imagery. A user might, for example, be interested in all of the significant earthquakes in Japan that generated a tsunami. If an earthquake generated a tsunami, the user could directly access the related information from the tsunami event table and display a list of locations that were affected by the tsunami as well as additional comments, references and photos. The data are also available through a web map interface[5]; see Fig. A–3).

A–2. THE TSUNAMI TRACE DATABASE (TOHOKU UNIVERSITY/ NRA)

Tsunami trace data of runup heights on past tsunamis are used to verify modelling techniques utilized for tsunami analysis and for bathymetry and topography models. However, many of the data items are based on ancient texts and other materials containing unverified information. Due to this fact, it is necessary to survey every single documentary origin of the trace records to clarify their reliability.

---

[3] http://dx.doi.org/10.7289/V5TD9V7K

[4] http://dx.doi.org/10.7289/V5JW8BSH

[5] https://www.ncei.noaa.gov/maps/hazards

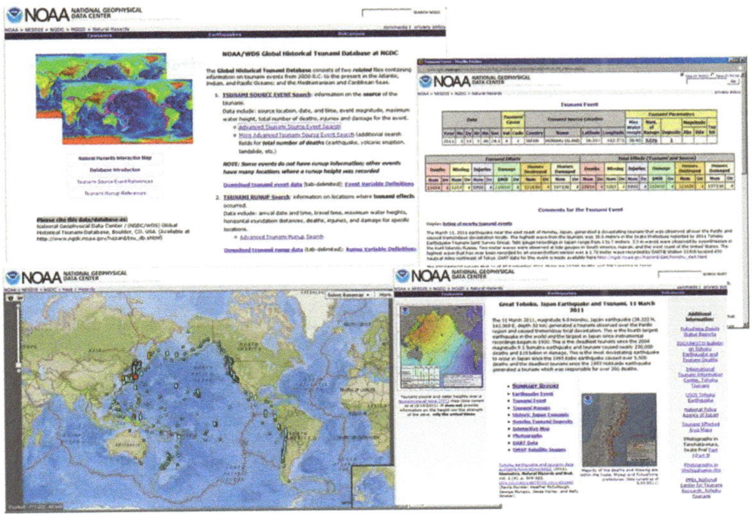

FIG. A–3. NCEI/WDS tsunami database web site and examples of data available online for the 2011 Great East Japan earthquake and tsunami, including an interactive map, tsunami travel time map and listing of tsunami event information.

To resolve the issues above, the Tsunami Trace Database, which takes reliability evaluation into account, has been prepared by Tohoku University and the Japan Nuclear Energy Safety Organization (JNES). Since it is critical to ensure fairness and objectivity in reliability evaluation, a review of the evaluation results has been conducted by the Tsunami Database Review Committee, which is composed of experts in historical tsunamis and tsunami engineering.

The database has information layers comprising tsunami information, literature information and trace information (see Fig. A–4), and it features a function to retrieve and read the information in each layer. As of 27 August 2024, the database covers the period from 684 CE to the present time, as shown in Table A–3. About 37 439 items of trace data (the total number of locations) were extracted from the collected literature (treatises, ancient texts and historical materials, and catalogues) and registered. Based on the acquisition and surveying of references, tsunami trace data over the past 500 years in Japan have been collected into a database. This information is then used to create about 50 data items for each trace (attribute information), such as the new and/or old geographical name, latitude and longitude, trace height (runup, inundation height, inundation depth, amplitude, etc.), measurement reference level for trace height and features measured.

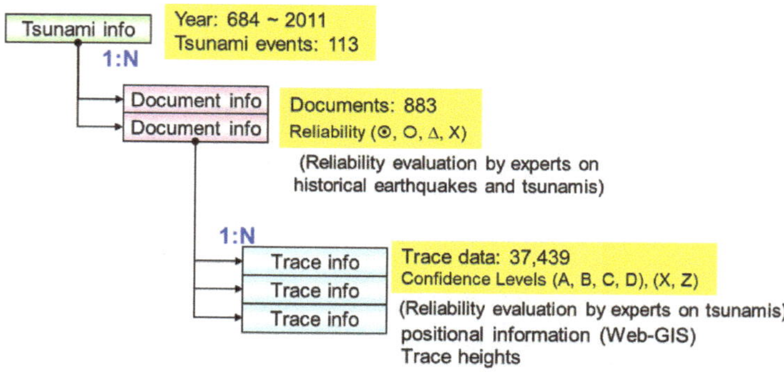

FIG. A–4. Association chart of information search (information layers and registered data).

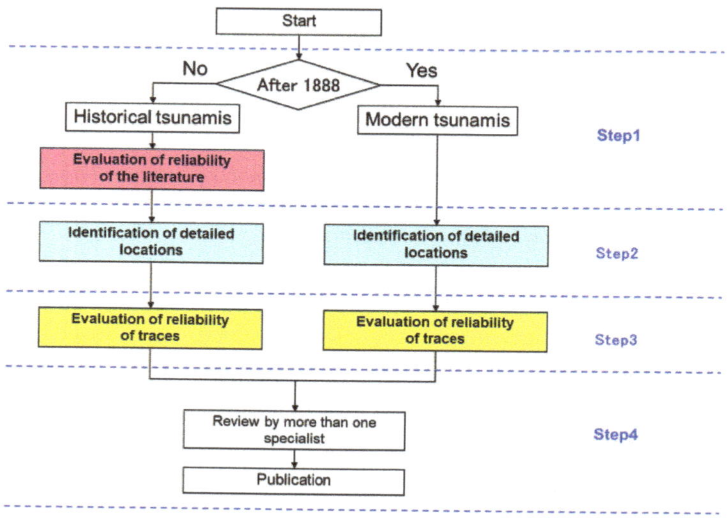

FIG. A–5. Procedure for examining each trace of historical tsunamis and modern tsunamis.

Each item of the tsunami trace data registered in the database is given a reliability rating based on judgement criteria, as shown in Table A–4. As a result, trace data can now accordingly be searched and filtered with reliability levels while it is utilized for verification of tsunami numerical analysis and estimation of tsunami sources.

The database can be accessed using a management system based on Web GIS (Geographical Information System) and enables Internet browsing of trace information placed over a base of seamless map images. The trace data are divided into historical and modern tsunamis from the perspective of reliability analysis, as shown in Fig. A–5. For historical tsunamis, the reliability of historical

TABLE A–3. LIST OF TSUNAMIS TARGETED FOR DATABASE REGISTRATION

Year and tsunami name	Tsunami magnitude			Number of pieces of trace data in database
	Abe's *Mt*	Imamura-Iida's *m*	Hatori's *m* [A-1]	
1498 Meio Tokai	8.5	3.0	3.0	296
1605 Keicho	8.2	3.0	3.0	214
1611 Keicho Sanriku	8.4	3.5	4.0	133
1677 Empo Sanriku	7.7	2.5	2.5	40
1677 Empo Boso	8.0	2.5	3.5	138
1703 Genroku	8.4	3.0	3.0	331
1707 Hoei	8.4	3.0	3.5	1392
1741 Kampo Volcanic (Oshima-Oshima)	8.4	3.0	3.5	338
1771 Yaeyama	8.5	4.0	4.0	275
1793 Kansei Sanriku	7.6	2.0	3.0	137
1833 Tempo (Off Yamagata Prefecture)	8.1	2.0	2.5	208
1843 Tempo Nemuro-oki	8.0	2.0	2.5	35
1854 Ansei Tokai	8.3	3.0	3.0	2269
1854 Ansei Nankai	8.3	4.0	3.5	913
1856 Ansei Sanriku (Off Hachinohe)	7.6	2.0	2.5	220
1894 Off the Southeast Coast of Nemuro Peninsula	8.2	2.0	2.5	86

TABLE A–3. LIST OF TSUNAMIS TARGETED FOR DATABASE REGISTRATION (cont.)

Year and tsunami name	Tsunami magnitude			Number of pieces of trace data in database
	Abe's *Mt*	Imamura-Iida's *m*	Hatori's *m* [A-1]	
1896 Meiji Sanriku	8.6	4.0	3.5	1074
1933 Showa Sanriku	8.3	3.0	3.0	2708
1944 Showa Tonankai	8.1	3.0	2.5	826
1946 Aleutian	9.3	4.0	3.0	31
1946 Showa Nankai	8.1	2.5	3.0	1891
1952 Tokachi-oki	8.2	2.0	2.5	392
1952 Kamchatka	9.0	3.0	4.0	224
1960 Chilean Tsunami	9.4	4.0	4.5	3008
1968 Tokachi-oki	8.2	2.0	2.5	1993
1983 Nihonkai-Chubu	8.1	2.5	3.0	4072
1993 Off the Southwest Coast of Hokkaido	8.1	3.0	3.0	2120
1994 Off the East Coast of Hokkaido	8.2	2.0	3.0	355
2003 Tokachi-oki	8.1	2.0	2.5	265
2006 Off the Chishima Islands	8.2	Not available	3.0	167
2010 Chilean Tsunami	Pending	Not available	3.5	400

TABLE A–3. LIST OF TSUNAMIS TARGETED FOR DATABASE REGISTRATION (cont.)

Year and tsunami name	Tsunami magnitude			Number of pieces of trace data in database
	Abe's *Mt*	Imamura-Iida's *m*	Hatori's *m* [A-1]	
2011 Off the Pacific Coast of Tohoku Earthquake and Tsunami	9.1	4.0	4.0	6487

* See Refs [A–2] and [A–3] for guidance on the tsunami magnitude scales. The definitions of Imamura-Iida's and Hatori's *m* are described in Section 2 of this Safety Report.

TABLE A–4. CLASSIFICATION OF CONFIDENCE LEVELS OF TSUNAMI TRACES

Before the 1960 Chilean tsunami			After the 1960 Chilean tsunami		
A	High confidence level	Trace information is described in ancient texts and local historical documents, etc., as well as the site can be confirmed at present. Furthermore, trace height has been surveyed and determined in recent years.	A	High confidence level	Clear traces with minor survey errors.
B	Moderate confidence level	Trace information is described in ancient texts and local historical documents, etc., as well as the site can be confirmed at present. However, trace height has not been resurveyed in recent years.	B	Moderate confidence level	Traces are obscure, but the surrounding conditions and witnesses indicate a reliable water level. Survey errors are insignificant.
C	Low confidence level	Trace information is described in ancient texts or told by word of mouth. However, the information is only limited to the name of regions and villages. The location of the trace cannot be tracked.	C	Low confidence level	Traces indicating waves having abnormally landed on a sandy beach, etc., or traces with significant survey errors due to survey points located away from the seashore.
D	Only as reference	Information speculated from relevant phenomena and descriptions of damage described in ancient texts.	D	Extremely low confidence level	Obscure traces overlapped by the effect of high water or typhoons, etc., or folklore or other ungrounded information.
X	Traces with least confidence	Traces with obvious errors in quotation and description. Traces that are not to be used or are to be excluded.			
Z	Traces that are to be judged based on the original literature used for catalogue creation	Traces classified as catalogues. Traces with values that can be summarized over relatively wide areas.			
Z	Duplication	Requotation from other literature.			
Z	Qualitative information	Qualitative information does not provide tsunami trace height but can be used to check the presence or absence of inundation.			

References: [blue]: Hatori [A–4], Earthquake Research Institute [A–5], Shuto and Unohana [A–6]; [pink]: Shuto and Unohana [A–7]; [green]: JSCE [A–8]; [yellow]: Iwabuchi et al. [A–9].

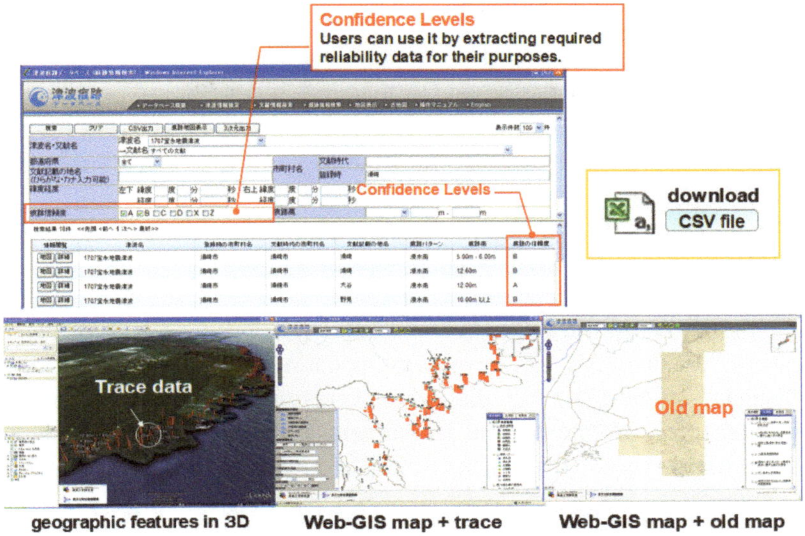

FIG. A–6. Example of the display of trace information search results. See https://irides.tohoku.ac.jp/eng/publication/database/tsunami-db.html.

materials was examined closely. To closely examine the locations described in the trace data, the accuracy of the data has been secured by a reference map at the time of each tsunami.

To provide the results of the research to local governments and residents living in coastal areas for tsunami disaster prevention, the database has been opened to the public and can be accessed via web based HTML forms. Figure A–6 shows an example of the display of trace information search results and the reference URL.

The coverage of target tsunamis in the database will be expanded, while investigations will also be continued in order to increase available trace records.

A–3. EUROPEAN COMMISSION PROJECT

A–3.1. Background information

As mentioned in Section A–1, the database of NGDC is one of the major sources of information about historical tsunamis, even in the Mediterranean region. However, other databases have been developed in Europe [A–10 to A–16]. Recently, Altinok [A–17] compiled historical documents in the Eastern Mediterranean for the period from 1410 BCE to 2011 CE with the inclusion of distant, local, volcanic and landslide generated tsunamis. Some of the main contributions of the Tsunami Risk and Strategies for the European Region (TRANSFER) project [A–18], funded by a grant from the European Commission, are (i) understanding tsunami processes in the Euro-Mediterranean region, (ii) understanding the tsunami hazard, vulnerability and risk assessment and (iii) identifying the best strategies for the reduction of the tsunami risk.

A–3.2. Database for the Euro-Mediterranean region

In the TRANSFER project, one of the work packages is dedicated to the tasks of (i) including new events or updating of existing events through palaeotsunami studies, (ii) including new events or updating of existing events of historical (pre-instrumental and instrumental) times, (iii) including sea level instrumental data, (iv) improving the format and items of the Genesis and Impact of Tsunamis on the European Coasts (GITEC) tsunami catalogue, (v) adding the 12 point tsunami intensity scale, (vi) improving and empowering of database architecture and (vii) integrating the Euro-Mediterranean tsunami catalogue into a worldwide tsunami catalogue. However, the inventory and characterization of the seismic and non-seismic tsunami sources are also evaluated in other work packages. Those work packages cover the tasks of (i) investigating tsunami genesis from historical earthquakes, (ii) investigating with the objective of reducing the uncertainties related to submarine seismic and non-seismic sources in selected areas and (iii) evaluating the potential of tsunami generation from seismic and non-seismic sources. Further information is available at TRANSFER [A–18].

In TRANSFER's deliverables, the Palaeotsunami Database is presented to collect data on tsunami inundations that occurred in the past. Evidence of palaeotsunamis is derived from coastal stratigraphy due to the presence of peculiar sediments or boulders. The dating of palaeotsunami deposits helps in correlating events with historical tsunamis or previous ones. This database provides two main types of information for developing tsunami scenarios and time dependent hazard calculations: locations of past inundations and their frequency. The data input in this database is included in the TRANSFER GIS Database and shared among project partners and database contributors. Each entry has a reference that should be quoted if the data are used.

This database is divided into two main parts: seismic sources of tsunamis in the Euro-Mediterranean region, and palaeotsunamis. GIS based data are stored in a TRANSFER mdb file, which includes detailed information under numerous subheadings.

The user can select each subheading from the database file TRANSFER.mdb, which will provide the requested data about the Mediterranean and North-East Atlantic region. In the Seismic Sources of Tsunamis in the

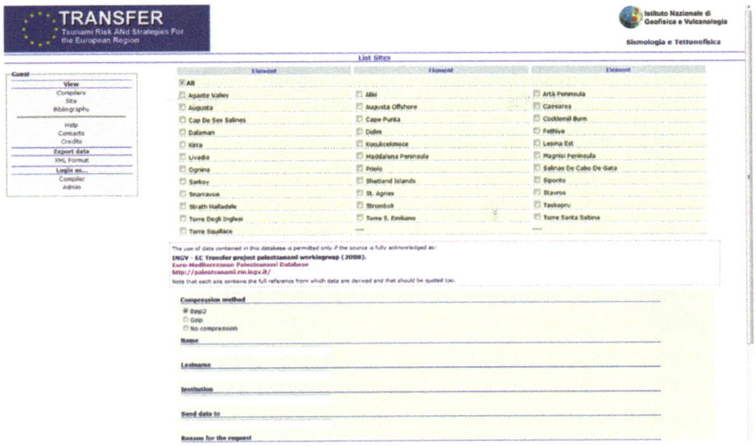

FIG. A–7. Web site of the TRANSFER project. Reproduced courtesy of the TRANSFER project [A–18].

Euro-Mediterranean Region database, three main parts are included: faults, individual sources and zones. Each part can be selected, and the necessary data can be obtained from the GIS file.

A–3.3. Data access and delivery

The data are available through the web site of the TRANSFER project (see Fig. A–7).

REFERENCES TO ANNEX

[A–1] HATORI, T., Classification of tsunami magnitude scale, Bull. Earthq. Res. Inst. Univ. Tokyo **61** (1986) 503–515 (in Japanese with English abstract), https://doi.org/10.15083/0000032863

[A–2] Intergovernmental Oceanographic Commission, Tsunami Glossary, IOC Technical Series 85, UNESCO, Paris (2019), https://unesdoc.unesco.org/ark:/48223/pf0000188226

[A–3] SATAKE, K., "Tsunamis", Treatise on Geophysics (Second Edition) Vol. 4, Elsevier (2015) 477–504, https://doi.org/10.1016/B978-0-444-53802-4.00086-5

[A–4] HATORI, T., On the Standard of Measurements and Observation Times, Report on the Chilean Tsunami of May 24, 1960, as Observed Along the Coast of Japan, Committee for Field Investigation of the Chilean Tsunami of 1960, Tokyo (1961) 165–165 (in Japanese),
https://dl.ndl.go.jp/pid/1379516

[A–5] HATORI, T., "Survey of tsunami", Handbook for a Field Survey in Areas Damaged by a Large Earthquake, Earthquake Research Institute, Tokyo (1983) 28–39 (in Japanese).

[A–6] SHUTO, N., UNOHANA, M., Field Surveys of Tsunami Heights from the 1983 Nihonkai-Chubu Earthquake Tsunami, Research Report of Experimental Station for Tsunami Disaster Prevention, Tohoku University (1984) 88–267 (in Japanese),
https://www.tsunami.irides.tohoku.ac.jp/hokusai3/J/publications/pdf/vol.1_8.pdf

[A–7] SHUTO, N., UNOHANA, M., Field Surveys of Tsunami Heights from the 1994 Hokkaido Toho-oki Earthquake Tsunami, Research Report of Tsunami Engineering, Tohoku University (1995) 1–156, Part 2 (in Japanese),
https://www.tsunami.irides.tohoku.ac.jp/hokusai3/J/publications/publications.html#12

[A–8] JAPAN SOCIETY OF CIVIL ENGINEERS, Tsunami Assessment Method for Nuclear Power Plants in Japan 2016, JSCE, Tokyo (2016),
https://committees.jsce.or.jp/ceofnp/node/140

[A–9] IWABUCHI, Y., et al., Development of tsunami trace database with reliability evaluation on Japan coasts, Doboku Gakkai-Shi, Ser. B2 Coastal Eng. **68** 2 (2012) 1326–1330 (in Japanese with English abstract),
https://doi.org/10.2208/kaigan.68.I_1326

[A–10] TINTI, S., MARAMAI, A., Catalogue of tsunamis generated in Italy and in Côte d'Azur, France: A step towards a unified catalogue of tsunamis in Europe, Ann. Geofis. **39** (1996) 1253–1299,
https://doi.org/10.4401/ag-4040

[A–11] GALANOPOULOS, A.G., Tsunamis observed on the coasts of Greece from antiquity to present time, Ann. Geofis. **13** (1960) 369–386.

[A–12] AMBRASEYS, N.N., Catalogue of tsunamis in the eastern Mediterranean from antiquity to present times, Ann. Geofis. **32** (1962) 113–130.

[A–13] PAPADOPOULOS, G.A., CHALKIS, B.J., Tsunamis observed in Greece and the surrounding area from antiquity to the present times, Mar. Geol. **56** 3–4 (1984) 309–317,
https://doi.org/10.1016/0025-3227(84)90022-7

[A–14] TINTI S., MARAMAI, A., GRAZIANI, L., The new catalogue of Italian tsunamis, Nat. Hazards **33** 3 (2004) 439–465,
https://doi.org/10.1023/B:NHAZ.0000048469.51059.65

[A–15] PAPADOPOULOS, G.A., FOKAEFS, A., Strong tsunamis in the Mediterranean Sea: A re-evaluation, Paper No. 463, ISET J. Earthq. Technol. **42** 4 (2005) 159–170,
https://www.researchgate.net/publication/255659860_Strong_tsunamis_in_the_mediterranean_sea_A_re-evaluation

[A–16] YOLSAL, S., TAYMAZ, T., YALCINER, A.C., Understanding tsunamis, potential source regions and tsunami-prone mechanisms in the Eastern Mediterranean, Spec. Geol. Soc. (London) **291** (2007) 201–230, https://doi.org/10.1144/sp291.10

[A–17] ALTINOK, Y., ALPAR, B., OZER, N., AYKURT, H., Revision of the tsunami catalogue affecting Turkish coasts and surrounding regions, Nat. Hazards Earth Syst. Sci. **11** (2011) 273–291, https://doi.org/10.5194/nhess-11-273-2011

[A–18] TINTI, S., et al., Tsunami Risk and Strategies for the European Region (TRANSFER), EC-FP6, Alma Mater Studiorum-Università di Bologna, https://cordis.europa.eu/project/id/37058

CONTRIBUTORS TO DRAFTING AND REVIEW

Abe, H.	Nuclear Regulation Authority, Japan
Azuma, K.	Nuclear Regulation Authority, Japan
Bugaev, E.G.	Federal Environmental, Industrial and Nuclear Supervision Service, Russian Federation
Chaytor, J.	United States Geological Survey, United States of America
Comerci, V.	Istituto Superiore per la Protezione e la Ricerca Ambientale, Italy
Costa, A.	Istituto Nazionale di Geofisica e Vulcanologia, Italy
Ebisawa, K.	Central Research Institute of Electric Power Industry, Japan
Fujii, N.	Tokyo Electric Power Services, Japan
Geist, E.	United States Geological Survey, United States of America
Ghosh, A.K.	Bhabha Atomic Research Centre, India
Godoy, A.R.	Private consultant, Argentina
Hebert, H.	Commissariat à l'énergie atomique et aux énergies alternatives, France
Henkel, F.O.	WOELFEL Beratende Ingenieure GmbH + Co. KG, Germany
Hibino, K.	International Atomic Energy Agency
Hussain, J.	Pakistan Atomic Energy Commission, Pakistan
Hyun, S.G.	Korea Atomic Energy Research Institute, Republic of Korea
Imamura, F.	Tohoku University, Japan
Iwabuchi, Y.	International Atomic Energy Agency

Jin, S.	Korea Institute of Nuclear Safety, Republic of Korea
Jones, H.	United States Nuclear Regulatory Commission, United States of America
Kânoglu, U.	Middle East Technical University, Türkiye
Khan, Z.A.	Pakistan Atomic Energy Commission, Pakistan
Kim, M.K.	Korea Atomic Energy Research Institute, Republic of Korea
Lynett, P.	University of Southern California, United States of America
Manna, P.D.	Istituto Superiore per la Protezione e la Ricerca, Italy
Matsuyama, M.	Central Research Institute of Electric Power Industry, Japan
Mian, M.T.	Pakistan Atomic Energy Commission, Pakistan
Mori, K.	Nuclear Regulation Authority, Japan
Nomura, S.	International Atomic Energy Agency
Park, S.H.	Korea Hydro & Nuclear Power Co., Republic of Korea
Pisharady, A.S.	Atomic Energy Regulatory Board, India
Rastogi, R.	Bhabha Atomic Research Centre, India
Roshan, A.D.	Atomic Energy Regulatory Board, India
Samaddar, S.K.	International Atomic Energy Agency
Satake, K.	University of Tokyo, Japan
Sato, T.	Nuclear Regulation Authority, Japan
Schindele, F.	Commissariat à l’énergie atomique et aux énergies alternatives, France
Sharma, P.K.	Bhabha Atomic Research Centre, India
Sugino, H.	Nuclear Regulation Authority, Japan
Suzuki, Y.	Japan Nuclear Safety Institute, Japan

Takao, M.	Tokyo Electric Power Company, Japan
Titov, V.	United States National Ocean and Atmospheric Administration, United States of America
Vasilievich, Y.V.	ROSENERGOATOM, Russian Federation
Viallet, E.	Electricité de France, France
Vittori, E.	Istituto Superiore per la Protezione e la Ricerca Ambientale, Italy
Watanabe, K.	International Atomic Energy Agency
Wu, C.	Nuclear Regulation Authority, Japan
Yalciner, A.C.	Middle East Technical University, Türkiye
Yanagisawa, H.	Tohoku Gakuin University, Japan
Yanagisawa, K.	Tokyo Electric Power Company, Japan

Consultants Meetings

Vienna, Austria: 18–21 July 2011
Tokyo, Japan: 28–30 November 2011, 3–5 October 2012
Rockville, Maryland, USA: 26–27 June 2012
Daejeon, Republic of Korea: 16–18 December 2012
Antalya, Türkiye: 2–4 May 2013
Gocek, Türkiye: 29 September–1 October 2013

IAEA
International Atomic Energy Agency

CONTACT IAEA PUBLISHING

Feedback on IAEA publications may be given via the on-line form available at:
www.iaea.org/publications/feedback

This form may also be used to report safety issues or environmental queries concerning IAEA publications.

Alternatively, contact IAEA Publishing:

Publishing Section
International Atomic Energy Agency
Vienna International Centre, PO Box 100, 1400 Vienna, Austria
Telephone: +43 1 2600 22529 or 22530
Email: sales.publications@iaea.org
www.iaea.org/publications

Priced and unpriced IAEA publications may be ordered directly from the IAEA.

ORDERING LOCALLY

Priced IAEA publications may be purchased from regional distributors and from major local booksellers.

24-04167E-T

MIX
Papier | Fördert gute Waldnutzung
FSC® C083411

Zeitfracht Medien GmbH
Ferdinand-Jühlke-Straße 7
99095 Erfurt, Deutschland
produktsicherheit@kolibri360.de